W9-BHU-482

GET UP AND GROW

LUCY
HUTCHINGS

Herb, vegetable and fruit growing projects for both indoors and outdoors, from She Grows Veg

GET UP AND GROW

Photography by
Philippa Langley

Hardie Grant
BOOKS

CONTENTS

SHEGROWSVEG

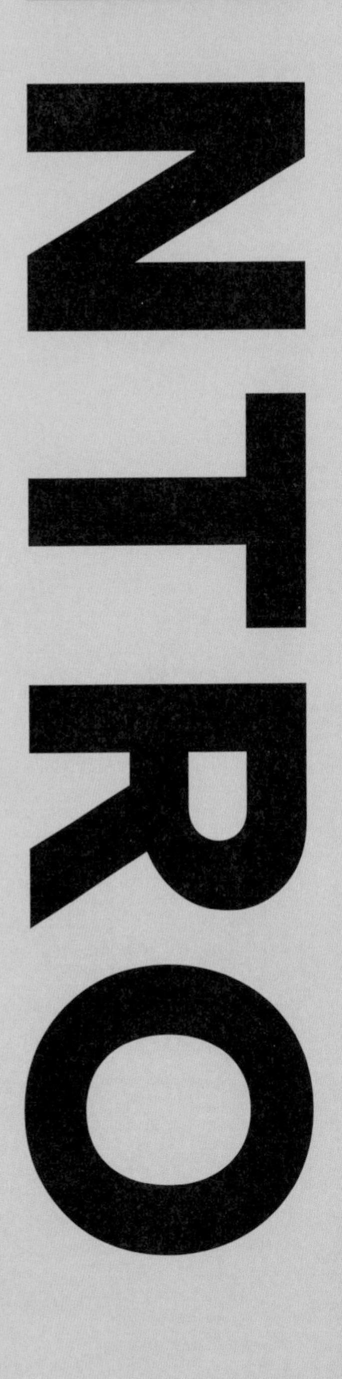

Happiness has always been my goal in life and in growing my own food I have felt the ultimate warm fuzzy glow.

It's no secret that growing and living with plants has a positive effect on mental and physical wellbeing. What could be more rewarding than nurturing a baby plant to adulthood and reaping the subsequent fruits of your labour? In a time of plastic-wrapped, intensively farmed, chemically laden produce, to have complete confidence in the provenance and security of your food must be the ultimate modern-day luxury. Nutrient levels in freshly picked food are substantially higher than shop-bought produce so we can say with confidence that growing your own food is good for the mind, the body and the heart.

My journey into the world of horticulture did not take, shall we say, the most well-trodden path. My first career was in fashion and for eight years I ran my own eponymous couture jewellery label. I was incredibly lucky and for many years lived a life that I had thought was my dream. Celebrities such as Cheryl Cole were fans of my work; I made stage outfits for artists like Kylie Minogue; my designs were featured in all the leading glossy magazines and my line was sold in some of the world's best-known stores. Dream job, right? Well, in theory, but as I came to a crunch point in my career, I faced the choice of taking the label to the next level or walking away. Crazy as it may sound, I had a moment of clarity and realised that I had been caught up in a whirlwind and that I hadn't noticed I was miserable. So, I did the unthinkable: I walked away. Like many creatives, the following years were filled with a diversity of projects, from multiple house renovations to art fairs, oh, and a couple of beautiful little girls came along! However, through it all – my years in fashion, getting to grips with motherhood, even my years at art school – there was one constant: I grew food. One way or another I always had something edible on the go, from herbs on the windowsill to lemon trees in my living room, from vertical herb gardens, to a windowbox vegetable patch on a fifteenth-floor balcony: growing food made me deeply and uncomplicatedly happy. So, when at the start of 2018 I was going through a deeply stressful life event and needing serious distraction, I turned again to my passion for growing edible plants and I can honestly say I have never looked back.

I started the @shegrowsveg Instagram account to engage with a younger global gardening community, never anticipating that it would gain the following and support that it quickly did. I retrained in horticulture and garden design, with the plan of becoming an edible garden designer. However, by the time I finished my training, the running of @shegrowsveg had become a full-time job and I have ended up with a career in garden media that grew organically. I am now privileged to champion heritage and heirloom vegetable varieties, charities and developers from around the world and am passionate about innovation in high-tech, future-proof food-growing. I am also lucky enough to visit some incredible projects around the world and to raise awareness of what can and is being done. However, I am never happier than when I'm in my greenhouse surrounded by a mind-boggling array of rare heirloom tomatoes which are the real stars of my Instagram account and blog.

Perhaps it is because I came in from left-field and because of my background in design, but I perceive there to be a large disconnect between ornamental gardening, houseplant-lovers and grow-your-own enthusiasts – and I don't get it! Aesthetics are important to me and I honestly believe you can have both beauty and abundant harvests; you can create the perfect botanical interior and reap a harvest from it too. Let's face it, the space for a designated kitchen garden is a serious luxury these days. Why should apartment dwellers, balcony gardeners or small garden owners be excluded from a taste of the good life just because of the traditional image of how food can and should be grown. It is time for a shake up, no more 'crops in pots' or cabbages in your flower beds, it is my aim with this book to show just what it is possible to create with edible plants. I have turned my designer's eye on the world of 'grow your own' and offer you a range of projects for all skill levels and for all spaces. Many of the projects are for indoors. A lack of outside space should not preclude you from enriching your life with home-grown food, plus indoor-growing offers the opportunity to break the rules and cheat the seasons. I will show you how you can grow your food beautifully and do it all year round – goodbye shrivelled market vegetables flown in from halfway around the world.

And that's another reason I am so passionate to share this book and hope to ignite a fire for self-sufficiency in you. So much of our food is flown in from other countries, giving it an epic carbon footprint. You've heard of the concept of food miles, the lower the food miles of your diet, the less impact your diet is having on this amazing planet we live on. The fact is that if we only ate what is seasonally available from our immediate environment the world would be a far healthier place as, almost certainly, would we. I don't pretend that this is a realistic lifestyle for most people, however being conscious of our food miles and how our food is produced will always steer us in a better direction. And what better way to instantly reduce our food miles than to start producing some of it at home?

As you move through this book you can pick and choose projects that will suit your home and your lifestyle. You'll find general guidance on growing, plus detailed care guidelines relevant to each project. You will also find a bit of sciencey stuff. Personally, I have always preferred to understand why I am doing something, not just to be told what I am supposed to do. It is easier to understand why something is done a certain way when you understand the cause and effect of your actions.

By the end of this book I hope you will be bursting with ideas and inspiration to create a truly jaw-droppingly beautiful, edible home to make you happier, healthier and feel like you are taking a step in the right direction for the planet. Whether you copy the projects exactly using the comprehensive suppliers' list (see page 150) to bring each project off the page and into your reality, or if the projects spark your creativity to take the concept and run with it, it is my honour to share my passion with you.

THE BASICS

GUTTER BOLTS
1¼ x 5/16 WHIT
5/16
5/16 x 3" W
5/16 x 2½ WHIT BOLTS
5/16 NUTS WHIT.
SET STUDS
1½ x ¼
1½ x 5/16
7/16 x 2
2 x 5/16
STUDS
SET STUDS
½ x 3/8 WHIT
½ x 1¾ SET STUDS
7/16 WHIT NUTS

The Basics: Potting up

Growing in smaller spaces and in the home almost always means growing your plants in pots or containers. It is vital to take time to give your plant's root environment some love – trust me, it will pay it back in harvests. Picture if you will, an astronaut bounding across the surface of the Moon, that iconic spacesuit is like a little oasis of life in an otherwise inhospitable environment. That suit must supply everything the astronaut needs to sustain the human body in one small space – air, warmth, water, waste processing and nourishment; outside that oasis the astronaut would die very quickly. Think of your plant's pot like a spacesuit, it allows that plant to grow outside its normal environment in the soil, and venture where no plant has gone before. Without that little oasis around its roots, just like the suitless moonwalker, your plant won't survive. You can see that spending a little time on creating the right root environment for your plant can pay dividends.

CHOOSING A POT

When choosing a pot there are two important factors to consider and sadly no, neither is how devastatingly beautiful it is. What really matters is drainage and size.

Drainage – There are essentially two types of pots and planters, those with drainage holes in the bottom and those without. Put simply, pots with holes are best used for plants growing outdoors and pots without holes are for indoor use. Why does it matter? Well, a pot with no drainage holes, when used outside, will quickly fill up with water with the first decent rain, becoming waterlogged and leading to potential root rot or even completely drowning your plant. Yes, plants can drown at the roots! I strongly advise against using no-drainage pots outdoors. Likewise, a pot with holes used indoors will merrily drain any excess water over your floor or furniture with each decent dose from the watering can. This problem is overcome by placing the pot on a saucer to catch excess or escaping water, a great solution as you can also easily see when a plant has sufficient water without having to worry about overwatering.

Size – Whether you are repotting something you have grown from seed or cutting yourself, or if you are potting on a recently bought plant, this rule applies: the new pot you transfer your plant to should be roughly 5 cm (2 in) larger in diameter than the original pot. This will be sufficient new room for the roots to grow while also giving a little boost of nutrition from new compost.

the particular needs of that plant, however multipurpose compost is a good basic media for most plants. But there is one compost rule I would implore you to follow, and that is to always use peat-free. Why does peat-free matter? Well, for millions of years our planet has done a really efficient job of locking away carbon dioxide (CO_2) in peat bogs, which cover roughly 3 per cent of the world's surface. Plants draw in and fix CO_2 from the atmosphere during photosynthesis (that's how plants make their own food) and the naturally waterlogged condition in bogs massively slows plant decompensation, creating deep layers of peat and effectively locking away the CO_2 in the decomposing plants where it can't do any harm. That is until humans realised what a fabulous, fertile resource peat bogs were, started to drain them for farmland and to harvest peat in large volumes for use as the perfect plant growing media. Damage to peatland is thought to account for up to 6 per cent of the carbon released into the atmosphere as a result of human action. Our making a commitment to use only peat-free compost is making a positive step to reduce our carbon footprint. There are already many fantastic peat-free products on the market, and the more demand there is for peat-free, the more time and funding will be invested in alternatives, resulting in more and better peat-free composts in the future.

GROWING MEDIA

Growing media is anything that surrounds the roots of a plant. In garden borders that is soil; in hydroponics it is often rockwool, expanded clay pebbles or coconut coir; in containers and indoor gardening this is mostly compost.

Compost – General-purpose compost will be fine for most plants. Speciality composts for cactus, citrus and bonsai are worth considering: they are widely available and tailor-made to support

Coir – Coir is made from the husks of coconuts and is widely available as compressed blocks that expand in water. Coir is easy to buy online from horticultural suppliers or check out your local pet shop as it is a main substrate in vivariums for reptiles and amphibians. The thing to note with coir is that it is not an alternative to compost. Though some nutrient-enriched coir products are available, coir in its pure form is sterile and has little to no nutritional value for plants, a plant cannot thrive in coir alone for long.

Expanded clay pebbles – Another hydroponic growing media, these can be very useful. Made of fired aerated clay, they are light, attractive and perfect as a mulch on the surface of unsightly compost in pots while also helping to deter soil-born pests. Additionally, they can be used to create a drainage layer in pots and are a lighter alternative to gravel and stones.

Sphagnum moss – Another fantastic soil disguiser, this moss acts like a natural sponge, trapping water and slowing the rate at which the compost dries out. It is widely available from horticultural and pet suppliers in compressed bocks.

POTTING UP FOR OUTDOORS

This is 'potting on', the practice of relocating a plant from one pot to a slightly bigger, new pot to give it more growing space. Start by filling the base of the new container with compost until it reaches a height where the top of the old plant pot is level with the top of the new container when sat inside. Add a few more scoops around the edge of the new container, so you have compost part way up the sides and a hollow in the middle. Place your hand over the top of the old plant pot containing the plant you wish to transplant, so you are covering the surface of the soil as much as possible. Turn the old pot upside down and press down on the bottom of the pot to encourage its contents to fall out onto your awaiting hand. (This is easier to do if the compost is relatively dry, so hold back on watering for a few days before potting on.) If the plant refuses to budge, squeeze the sides of the pot in a few places to loosen the roots' steely grip. Resist the temptation to grab the stem of the plant and yank it out, as this can snap the roots or stem, damaging or killing your plant. Once you have extracted the contents of the old pot, place it into the middle of the new pot and top up the compost around the edges. Water liberally until you see it trickling out of the bottom of the container and voila, plant relocation complete!

POTTING UP FOR INDOORS

When potting up for indoors, focus seriously on suitable drainage. As previously discussed, one option is to use a pot with drainage holes placed on a plant saucer. Another solution is to look for 'self-watering' pots that manage their own drainage: they come with an integrated drainage area at the bottom that also acts as a reservoir, watering from the bottom up as required. However most indoor planters are simple pots without holes and will require you to create your own drainage layer in one of two ways.

Method one: Leave your plant in its rather unattractive plastic plant pot and place that inside the ornamental container you wish to use: there will be a space between the bottom of the plant pot and the base of the container where excess water can collect. This is my recommended method, as it means that you can easily lift and check the water situation beneath, plus it allows you to re-style your plants and their containers to your heart's content!

Method two: If you want to install a plant directly into an ornamental plant pot it is possible, though this method includes a higher margin for error as it relies on guessing what is going on below the compost surface. Use expanded clay pebbles to create a drainage layer in the bottom of the container. The pebbles need to be at least 3 cm (1 in) deep, this will create an area into which excess water can drain. Then place your compost and plant on top of the pebbles as per the method for outdoor potting on.

5/44

The Basics: Light

Light is vital to plants; without it they cannot photosynthesise to make food and will have insufficient energy to grow. Understanding and maximising the available light in and around your home will result in happy, healthy plants and hopefully beautiful, abundant harvests. Very roughly below are the expected light levels from the windows in your home:

NORTHERN HEMISPHERE

North-facing window – offers the lowest light levels as it receives no direct sunlight at any time of day.

South-facing window – the highest levels of light throughout the day with six or more hours of direct sunlight.

East-facing window – bright indirect light for most of the day with direct light in the morning. Morning sunlight is not as strong as afternoon sunlight so, though it is a bright spot it has lower light levels than a west-facing aspect.

West-facing window – high light levels with bright indirect light most of the day and direct sunlight through the afternoon when the sun is at its most powerful.

SOUTHERN HEMISPHERE

North-facing window – the highest levels of light throughout the day with six or more hours of direct sunlight.

South-facing window – offers the lowest light levels as it receives no direct sunlight at any time of day.

East-facing window – high light levels with bright indirect light most of the day and direct sunlight through the afternoon when the sun is at its most powerful.

West-facing window – bright indirect light for most of the day with direct light in the mornings. Morning sunlight is not as strong as afternoon sunlight so, though it is a bright spot it has lower light levels than an east-facing aspect.

Though the guidelines seem straightforward you need to temper them with the local environment. Obstructions both natural and manmade can alter the available light levels in your windows dramatically. Trees and buildings can obstruct direct sunlight and alter the expected light level of a window. Additionally, height can make a huge difference, the south-facing window of a tenth-floor apartment will offer dramatically more light than the south-facing window of a basement. The key is to get to know YOUR growing environment. Watch how the light moves through the rooms of your home. Where sharp shadows are cast on the walls and floor you know you are getting direct sunlight; if you have a huge building across from your window, observe if the sun peeks round it at any point during the day. By taking a few days to assess the available light in your home, rather than assuming its theoretical light, you will have significantly happier plants.

KNOW THE RULES AND HOW TO BREAK THEM!

Plants generally grow in spring and summer. Annual plants live for one year and have to do all their growing, flowering and setting seed in one growing season, after which they will die. Perennial plants can live for many years but will still put on their main growth and flower and fruit within the same period as annuals. But why? What is it about spring and summer that stimulates plants into this incredible burst of life and productivity? Well, it comes down to two simple factors: heat and light. Plant functions such as photosynthesis are at their optimum between 25–36°C (77–97°F), but either side of this range plant growth slows down. As we know, light is necessary for plants to fuel themselves. Unless you live near the equator, in winter, the days are shorter and there are simply not enough hours of daylight available for plants to make sufficient energy to sustain active growth. As a result, plants enter a period of dormancy while they wait for daylight hours to lengthen sufficiently to stimulate growth again. When heat and light levels hit the sweet spot, usually around April in the Northern Hemisphere and October in the Southern Hemisphere, dormancy is broken and off they go again. Once you understand how and why plants grow in seasons, you also start to understand that it is possible for us to break the rules and grow outside the seasonal confines. This is great if you are growing food and want to keep the good times going throughout the year. Plants may have evolved to only grow through half the year, but we need to eat year-round. Growing in your home is the perfect way to break from seasonal growing and manipulate the environment to sustain year-round food production.

Growing in your home automatically removes the issue of heat. Most homes are heated in winter and as long as this internal temperature is maintained to at least 18°C (64°F) and ideally 20°C (68°F) or above, your plants will continue to grow. If you are concerned that you cannot maintain these temperatures, placing plants on a heat mat (available cheaply online) can also solve the problem.

Conquering the issue of light requires fractionally more effort, though is far from hard to master and well worth doing properly. Though our homes are filled with light, standard electric light bulbs do not supply the same wavelengths of light as sunlight. Stick a plant under a standard lamp and it will still grow towards the nearest window in an effort to seek out a strong light source, as the lamp light is not enough to satisfy a growing plant's needs.

In order to properly supply the correct levels of light artificially we need to use grow lights. These come in many forms from clip-on table lamps, to single bulbs fitted to a normal lamp, to huge strip lights for larger rooms. There is something for every space, environment and budget, and they can be purchased online. There is one very important phrase to look out for, whatever light you decide go with, it must be 'full spectrum'. Normal white light is made up of lots of different colours which have different wavelengths. Plants utilise many of these to grow but, the most important ones are blue light, which stimulates leafy growth making strong, sturdy plants, and red light, which controls flowering, vital for fruit. Only a full spectrum light will supply enough of both wavelengths to support your plants. As a designer I should point out that a full spectrum bulb normally casts a pink light which may not be desirable in your home. If you want a white light you will need to look for a light sold as 'full spectrum sunlight'. These are also widely available but may require more research: reptile supply shops are a good source but be sure to purchase a strong enough wattage, many reptiles are shade-dwellers and so lights of varying strength are sold accordingly. If in doubt speak to a staff member and say that the light needs to support plant growth. It is important to fully arm yourself with all the information before you make your purchase, as the right grow lights will last you many years and will be a game-changer for your life as a plant parent!

The Basics: Feeding

Anyone who has seen a growing toddler or teen will know that growth requires FOOD! The same is true for plants. Plants do their growing in spring and summer and during this time they will certainly benefit from, if not desperately require, a good regular dose of feed. Many nutrients are needed to sustain healthy plant growth, the three main elements being Nitrogen (N), to promote healthy green growth of leaves and stems; Potassium (K) which supports development of flowers and fruit; and Phosphorous (P) which supports healthy roots. Most general-purpose feeds will offer a good balance of these three nutrients, along with a whole host of other important trace elements and minerals that your plants need. Additionally, there are specialist feeds available for species such as citrus and orchids which offer a specifically tailored set of nutrients.

I would strongly advise you choose an organic-based feed such as seaweed. Gory as it may seem, bone meal, fish, horn and blood are also commonly used organic feeds though they are usually sold in a slow-release granular form. What is the point of growing your own fabulous food, food you can be confident has never seen a drop of pesticide or herbicide, but then treating it to a diet of synthetic nutrients from petrochemicals and mined minerals? Let's keep it natural guys, our bodies are a temple. My favourite plant food is a byproduct of a renewable energy plant, that is also vegan-friendly – should the idea of non-vegetarian vegetable plants not appeal (see suppliers' page 150).

You may think that with all these fabulous, growth-promoting nutrients to offer, playing fast and loose with the plant food would be okay? But hang on! Overfeeding a plant can have more disastrous effects than giving no food at all. An underfed plant will likely slow or stop growing, an overfed plant can quickly die, poisoned by an overdose of the very minerals that promote growth. There are two key tricks to nailing feeding. Firstly, always follow the manufacturers' instructions on your selected food as to how and when to administer it. There is no hard and fast rule to feed strength and regularity, each feed is nutritionally unique and will come with instructions specific for best practice. Secondly, feed during the growing period, that's basically April through to September in the Northern Hemisphere and October to March in the Southern Hemisphere. If you are growing under artificial light, feed through the period of flowering and fruiting as this is your artificial 'summer'.

The Basics: Watering

Even a novice grower knows that plants need water to grow – they really, really need it and like all living things, without this precious resource plants soon die. However, it is possible to have too much of a good thing. Soil that is consistently too damp can act as the perfect breeding ground for disease and fungal infection, destroying your plant from the roots up, as well as effectively causing it to drown. Plant are amazing lifeforms, busy every second of the day and night with incredible, invisible processes that frankly boggle the mind. They draw in carbon dioxide (CO_2) through their leaves and stems in the sunlight and give off oxygen in the process of photosynthesis – it's how all plants create sugars to feed themselves and is what makes them such incredible air purifiers. However, they also breath oxygen through every part of themselves, including the roots. It's called respiration and it is how they turn the sugars they made through photosynthesis into the energy they need to grow. There are loads of tiny air pockets in soil, containing a perfect little oxygen supply for plants. When this soil becomes waterlogged these tiny pockets fill up with water and there is no air left for the roots to breathe, so the plant will literally drown. You can see why it is so important to attune to the water requirements of your plant babies.

When growing indoors you are totally responsible for all your plants' water needs. Even outdoors when rain can share the responsibility, you will find that plants in containers dry out much faster than those in the ground. There's no magic formula for the perfect watering regime, as just like people, every plant uses water at a different rate. Additionally, differences in position such as proximity to direct sunlight or the varying warmth of different rooms will affect the speed at which containers dry out. That is why it is important to let the soil tell you when it is time to water. As it dries, compost changes colour, turning from a rich dark brown/black to a much lighter brown/grey when dry. But the soil's surface can deceive you and is not necessarily giving an accurate impression of what lies beneath – many windowsill pots will have a surface baked dry by the sun while the compost beneath is perfectly moist.

So how do we know when to water? Easy, you have to get your hands a little dirty, or at least one finger. Stick a single finger down several centimetres (an inch or so) into the soil. If your finger comes out dry with dusty compost particles that easily brush off, it is time to water. If your finger is slightly clammy and the compost clings to it, the moisture levels in that container are fine. By forming the habit of watering via observation you will always take the cue from your plants and give them what they need, confident you are watering rather than overwatering.

Trouble Shooting

Plants are living organisms, so growing them is subject to a margin of error, but forewarned is forearmed. If you encounter issues or things are not going to plan, don't panic. It happens to everyone all the time, and being able to identify a problem and remedy it quickly is an important skill to learn.

PESTS

Wouldn't it be lovely if growing indoors meant no issues with pests? You might hope that wildlife would stay outside, but plant pests have an uncanny ability to appear from nowhere. The way I see it, bugs are part of the natural order of things – if you have a few pests chowing down on your plants it is because you have created a perfect ecosystem for them. I take it as a compliment, but I won't let a few bugs become a major infestation that ruins my harvests: I want to be the one eating my plants. Luckily there are many options available for both indoors and out, both to treat a surprise infestation and to manage their long-term prevention – just don't reach for the chemicals. Remember these plants are food and what goes into them goes into you. Organic pest control is diverse and fascinating, often tailored to a specific pest. Some of the best methods involve introducing a tiny predator to the prey on the pest. Even minute organisms can effectively wipe out a pest population despite being invisible to the eye. Just remember to give your harvests a good rinse before eating.

Aphids – It's pretty easy to see if these sap-sucking green, black or sometimes pink bugs alight on your plant babies. They are several millimetres (⅛ in) long and you'll see them cluster on the tender tips and buds as well as undersides of leaves. Ladybirds and lacewing larvae are fantastic at controlling these irritating aphids in the greenhouse. However, would you want to release an army of flying creatures into your home to control an indoor infestation? A spray containing horticultural soap (neem oil) is a good way to kill aphids without harmful chemicals. The fatty acids in the soap clog aphids' breathing holes and moving parts, and the soap isn't a poison so it's safe for edible plants. Though honestly, one of the best ways to reduce an aphid population is manual control, yup, I mean squashing the little beasties. Running your fingers over the leaf undersides and tender tips can take them out in their hundreds and is darkly satisfying.

Sciarid fly (fungus gnat) – These irritating little flies seem to materialise from thin air, one day none, and the next hundreds buzzing about. Unfortunately their eggs are present in most compost and are stimulated to hatch in the warmth of your home, so they are somewhat inevitable. They are an irritation rather than a problem, except for during propagation when the tiny maggots can devour a seed before it germinates. You can control these little suckers at all stages of their life and a combined attack is best. Firstly, yellow sticky traps will reduce the flying adults. Predatory nematodes (see suppliers' page 150) will deal with the larval stage: just water the microscopic worms into the soil. These nematodes are safe to humans and animals but will attack the maggots that have hatched. Repeat the treatment after two weeks to kill maggots that have hatched since the first treatment. There is also a predatory mite called *Hypoaspis* that you sprinkle on the compost. This hungry little thing will munch on any eggs and larvae, and will survive for a good while without food (meaning they guard against future attacks). For a final bullet in your gun, try watering your plants from the bottom up – pouring water into the base of the container to be absorbed upwards. Sciarid fly prefer a moist environment so keeping the surface of the compost dry may deter them from laying eggs.

Thrips – These sap-suckers are around 2mm (½ in) long, either black or cream, and collect on the undersides of leaves. They can be tricky to spot – unlike the damage they cause which is usually silvered foliage and small black spots. The good news is that the same predatory nematode that attacks sciarid fly also preys on thrips, so you can kill two birds with one stone. Sticky traps also work but interestingly thrips are more attracted to blue rather than yellow ones.

Red spider mite – Though invisible to the naked eye, red spider mite can cause a staggering amount of damage if left unchecked. Look for a lighter-coloured stippling effect on leaves showing that the tiny bugs are draining the plant sap from beneath. As an infestation progresses you'll see fine webs at the apexes of leaves, which will start to cover more leaves and stems, the plant will become yellow, drop its leaves and be close to death. Treat infestations early to prevent the spread of these horrible little beasties. An initial spray with horticultural soap will start to control the numbers, but do this before introducing predators or you will wipe them out too. The predatory mite *Phytoseiulus persimilis* will munch its way through the remaining population. Annoyingly, once you've had spider mite, it's hard to avoid repeat attacks: these little bugs hide in the smallest crannies to emerge, breed and reappear. Get in early with a pre-emptive strike in spring by introducing *Amblyseius andersoni*, a predatory mite that will devour spider mites as they appear early in the season.

Mealy bug – These easy-to-spot sap-suckers look like tiny balls of cotton wool on the leaves and stems of your plants. Horticultural soap is the best means of attack, though that fluffy coating means spraying can be ineffective and it is often easier to wipe them off using the correctly diluted horticultural soap and a paper towel.

Plants that turn the tables on bugs

Carnivorous plants are fascinating, aren't they? At the bottom of almost every food chain on earth you find plants consumed by herbivores which are in turn eaten by carnivorous predatory animals. There is something so wonderfully ironic about a plant that turns the tables and preys upon animals. We are most familiar with the visually striking pitcher plants and iconic Venus fly trap. Sadly, as amazing as they look, they are more likely to be attacked by plant pests than to consume any. But they will catch the odd housefly – and I like to keep a few around for that reason.

A couple of more diminutive and less showy carnivorous plants are, however, fantastic pest controllers. The butterworts (*Pinguicula*) with their sticky glandular leaves, and sundews (*Drosera*), covered in glue-like dew-drops, both function as excellent living fly paper. They are particularly effective at trapping sciarid flies and thrips, and many top gardeners use them to keep numbers down. Generally small and discreet, they can be planted within projects and growing spaces as ongoing pest control. Oh, how I love a useful plant!

recommendations for strength and regularity of feeding.

Pot-bound plants – to pot or not
When plants are grown in pots, there is only so far that their roots can go. If plants aren't regularly potted on their roots will start to circle the interior of the pot, becoming denser and leaving less room for nutrient-rich compost. Signs of a pot-bound plant include: roots escaping through drainage holes or from the top of the pot; plastic containers becoming distorted by tightly packed roots; growth slowing or stopping; the plant seeming thirsty and quick to wilt after watering; and the leaves browning at the edges.

When repotting a root-bound plant into a larger pot, use your hand to loosen the tightly packed roots so that they can spread out again in their new home. If they can't be prised apart, you will have to slit them, top to bottom in four places around the side, and in a big X across the bottom. Don't worry, as brutal as this may seem, this will not damage your plant but will give the roots a chance to escape the stranglehold of the outer roots.

There is an argument for not potting up your plants. Some plants grown indoors have the potential to become giants: a mature lemon tree can reach 4–8 m (13–26 ft) high and a loquat is just as ambitious. An indoor plant of these proportions is just not practical so constraining the roots in a pot is one way to manage their size. The bonsai technique employs root-pruning to keep mature plants in small containers and yet remain healthy. To try this technique, use scissors or secateurs to remove some of the larger roots from the

OTHER ISSUES

Mould and mildew – The damp conditions found at the bases of plants are a perfect breeding ground for mould and mildews. These can be fluffy black, white or grey coatings on the plant or soil, or white powdery coatings on leaves. Prevention is the best route, and you can start with good hygiene. Fungal spores are common in old compost so avoid reusing it and sterilise it (by pouring boiling water over it) if you can't. Make sure not to overwater plants to avoid root rot and clear away dropped leaves or dead plant material before it decomposes. If you see signs of mould on the soil surface or on the plant remove these parts if you can and spray with strong chamomile tea. Chamomile contains natural anti-fungal properties and it will start to keep things under control.

Nutrient deficiency and overdose – Each nutrient deficiency will cause slightly different symptoms, but most will manifest as a discolouration of the leaves. Signs to look for are the leaves becoming darker or lighter, or yellowing of the whole leaf or just the veins. As most liquid feeds offer a balance of all necessary plant nutrients, a dose of the appropriately diluted feed should resolve the problem no matter which element is deficient.

Overfeeding or using a feed that is too strong actually burns the roots of the plant, leaving it unable to absorb any nutrients from the soil. Ironically it will often result in death, as the plant effectively starves to death despite the high level of nutrients in the root zone. Ensure that you follow the manufacturers' outside and base of the rootball. This will make space to add fresh compost to the current container and for the roots to expand despite remaining in the same-sized pot.

Hand-pollination – So, for weeks you have diligently watered, you have followed the feeding schedule to the letter, you have vanquished every hint of a pest, you have the happiest, healthiest plants in existence – yet the flowers keep dropping off without setting fruit. The problem is likely to be one of pollination and it may be time to get hands-on with the birds and the bees, specifically the bees.

Plants growing outside will naturally be pollinated by flying insects but when growing indoors this is unlikely to occur. So your plant babies' flowers may need help to achieve pollen transfer. With tomatoes, (bell) peppers, chillies and aubergines (eggplant) you can gently tap each of the flowers – this should be enough to shake things up and enable flowers to self-pollinate. On most fruits including citrus, take a cotton bud or small clean paintbrush. Gently tickle the stamens inside each flower: pollen will build up on the bud (or brush) and be transferred, flower to flower, and when you are doing this you are literally playing bee.

PLANT DIRECTORY

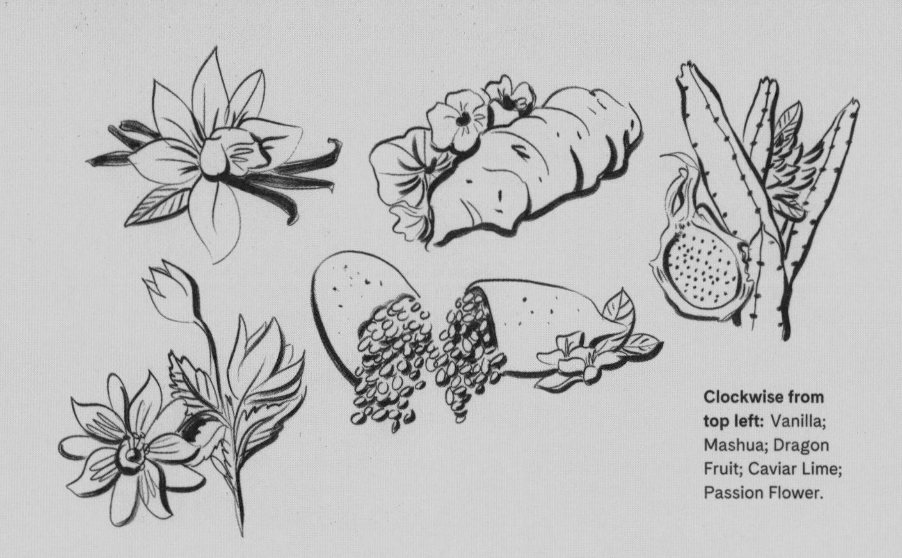

Clockwise from top left: Vanilla; Mashua; Dragon Fruit; Caviar Lime; Passion Flower.

Conditions:
Each plant notes the optimal light, temperature and moisture required.

Fruit

ORNAMENTAL PASSION FRUIT/ EDIBLE PASSION FRUIT
(*Passiflora caerulea/ Passiflora edulis*)
Full or partial sun. Warm. Average

Ornamental passion fruit vines *(P. caerulea)* are widely available and though often considered inedible this is not true. Though not as large, juicy or sumptuous as edible passion fruit *(P. edulis)*, the fruit produced is perfectly edible as well as sporting some beautiful flowers. *P. edulis* is a little harder to get hold of in temperate zones, because it is a heat-lover and not as suited to temperate gardens as its ornamental cousin. It is possible to find however (see suppliers' page 150) and will make a stunning houseplant that will also provide you masses of delicious fruit.

LOQUAT

(*Eriobotrya japonica*)
Full or partial sun.
Temperate. Average

Also called the Japanese plum, it is grown for both the fruit and the leaves which can be made into a tea. This evergreen tree can grow up to 10 m (33 ft) tall but can be kept more manageable if grown in a container. Aim for a self-fertile variety such as 'Gold Nugget' or 'Mogi'. It is a sculptural plant with large ridged, velvety leaves. Unusually for a stone fruit it flowers in autumn and fruit is ready for harvest in spring and early summer. Fruit is best eaten fresh when dark orange, tastes similar to apricots or mango, and is particularly good in preserves.

—

DWARF BANANA

(*Musa acuminata* 'Dwarf Cavendish', *M. paradisiaca* 'Dwarf Orinoco' *and M.* 'Brazilian')
Full or partial sun. Warm. Moist

Dwarf bananas are already a popular houseplant but selecting the right variety will also provide an edible crop of bananas. 'Dwarf Cavendish' is widely available as an ornamental houseplant and produces edible bananas for cooking with. For a better-tasting option find 'Dwarf Orinoco' or 'Dwarf Brazilian'.

—

LYCHEE

(*Litchi chinensis* 'Kwai Mai Pink')
Full or partial sun.
Temperate. Average

This is a hardy lychee variety and will need to be kept in a sheltered position outside for at least a month during winter as it is triggered to flower by winter temperatures below 12°C (54°F); however it is not a fan of frost, so an outside greenhouse or sheltered balcony is perfect.

MANGO

(*Mangifera indica* 'Gomera 1')
Full or partial sun. Warm. Average

This is a hardy mango variety and though it could be years before it matures enough to produce fruit, in the meantime it will make a beautiful plant for your home, with elegant long leathery leaves that are dark red when they first emerge.

—

DWARF POMEGRANATE

(*Punica granatum* var. *nana*)
Full or partial sun. Warm. Average

A compact tree form of the full-sized pomegranate, this will reach a maximum of 1 m (40 in) tall. However it will still put on the same dazzling display of vibrant red flowers all through summer, followed by full-sized fruit.

—

PINEAPPLE

(*Ananas comosus*)
Full or partial sun. Warm. Average

This is a member of the bromeliad family and is a popular choice for houseplant lovers. Once the main fruit has been harvested, the plant will send out side shoots which can either be removed and rooted to form new plants or left on the parent plant where they will grow new fruit. Pineapples also grow babies around their base, called suckers – these can be carefully removed to form new plants. Additionally, the leafy top of the fruit can be encouraged to root and become a new plant .

CITRUS **which includes lemon, orange, lime, grapefruit, calamondin, yuzu, ginger lime**

(*Citrus x limon, C. sinensis, C. x aurantiifolia, C. x paradisi, C. x microcarpa, C. x australasica and C. x junos*)
Full or partial sun. Warm. Average

Although citrus trees are hardier than many people believe, they love heat and will grow well indoors though will need to be misted regularly to keep humidity high. In spring and summer, they erupt in highly scented white flowers that will strongly perfume a whole room. Once set, the fruit will take a full year to ripen. Feed regularly from March to October with a specialist citrus feed high in nitrogen.

—

DRAGON FRUIT

(*Hylocereus purpusii/ Hylocereus undatus*)
Full or partial sun. Warm. Average

Unlike most cactus, the dragon fruit cactus is from the tropics and therefore likes to be kept relatively moist. In its native habitat it is a parasitic plant that will grow on the branches of host trees. It is therefore necessary to give this plant something, ideally wooden, to climb. Dragon fruit bloom at night and cross-species pollination is the best way to ensure fruit set, meaning you need two plants. Certain varieties pollinate other varieties best, the two I have listed here are the dragon fruit most commonly cultivated in agriculture, and its best friend for pollination. They are also really easy to propagate from cuttings so one can become many really easily, just place in moist compost and keep somewhere warm.

PRICKLY PEAR

(*Opuntia ficus-indica*)
Full or partial sun. Warm. Dry

This classic pad cactus will produce oval fruits on the tips of its pads after a glorious display of bright orange flowers. Pick fruit when it is green to yellow as it will be overripe when it turns red. Remember to wear gloves and carefully brush off all any hair-like spines, though spineless cultivars are available.

—

KIWI

(*Actinidia deliciosa* 'Jenny')
Full or partial sun.
Temperate. Average

Kiwis have fuzzy-all-over fruits, and are phenomenally productive fruiters once established. They are climbers but will be as happy to climb the banisters as a trellis. Not all kiwis are self-fertile, but 'Jenny' is, meaning you will only need to grow the one plant to achieve fruit.

—

KIWI BERRY

(*Actinidia arguta* 'Issai')
Full or partial sun.
Temperate. Average

Unlike its cousin 'Jenny', the kiwi berry or hardy kiwi is devoid of fuzz and smaller in size. It is also self-fertile, and a productive fruiter once established.

Vegetables

TOMATOES

(*Solanum lycopersicum*)
Full or partial sun. Warm. Moist

Tomatoes have four growth habits and which you choose will be determined by where you can grow it.

Cordon or **indeterminate** tomatoes grow tall, reaching 2.75 m (9 ft) or more if unchecked.
Bush or **determinate** variates are smaller, growing to around 1 m (48 in) high but can have multiple stems and will sprawl without support.
Patio or **micro dwarf** varieties will remain at a very manageable 30–40 cm (12–16 in) height.
Tumbling varieties will, as the name suggests, have a cascading growth habit but will remain manageably small.

There are more than 10,000 varieties of tomato to try. Personally, I am devoted to heirloom varieties but many people swear by F1 or hybridised options. Whatever takes your fancy, patio and tumbling varieties are the best choice for small-space growing.

—

AUBERGINE (EGGPLANT)

(*Solanum melongena*)
Full or partial sun. Warm. Average

These plants can get bigger than you might expect. When growing indoors aim for a patio variety such as 'Patio Baby' which will remain an amenable size or use pruning to achieve a smaller multi-branched plant.

—

(BELL) PEPPERS/CHILLIES

(*Capsicum annuum/ Capsicum frutescens*)
Full or partial sun. Warm.
Average to dry

Highly adaptable and ornamental, they make great houseplants and can be cultivated using bonsai techniques to create manageable or even miniature plants.

CUCUMBER

(*Cucumis sativus*)
Full or partial sun. Warm. Moist

Cucumber vines can get very large, however patio varieties remain much more manageable when growing in small spaces.

—

SQUASH

(*Cucurbita pepo*)
Full or partial sun. Warm. Moist

When watering squash make sure to water the soil not the leaves to help keep powdery mildew under control.

—

CHARD

(*Beta vulgaris*)
Partial shade. Cool. Average to dry

A highly decorative and easy-to-grow leafy green with stalks in bright rainbow hues.

—

CELERY

(*Apium graveolens*)
Partial shade. Cool. Moist

A kitchen staple and a sculptural plant in its own right with lush leafy tops.

—

LETTUCE

(*Lactuca sativa*)
Partial sun. Cool. Average

Myriad varieties are available including highly ornamental varieties with scarlet tipped or deepest red foliage.

ROCKET (ARUGULA)

(*Eruca vesicaria* subsp. *sativa*)
Partial sun. Cool. Average to moist

A delicious but slightly tricky crop to grow, if unhappy it can be prone to 'bolting' when premature flowering ruins the peppery flavour of the leaves.

—

CLIMBING AND DWARF BEAN

(*Phaseolus vulgaris*)
Full or partial sun. Temperate. Average

Climbing beans come in green, red, pink, purple and yellow and will crop prolifically over a long season if you continue to pick them. Dwarf varieties will only reach around 40 cm (16 in) tall and will not require any support.

—

RUNNER BEAN

(*Phaseolus coccineus*)
Full or partial sun. Warm. Average

Though grown as an annual in temperate regions these are in fact a perennial in tropical climates where they originate. When you see the stunning floral display put on by this versatile bean it is easy to see why these were originally imported to Europe as an ornamental plant. 'Sunset' with its stunning peach flowers is my personal favourite. Keep picking to ensure cropping over a long season.

—

MASHUA

(*Tropaeolum tuberosum*)
Full or partial sun. Warm. Average

Every part of this perennial nasturtium climber is edible from the tubers to the leaves and flowers which are particularly beautiful and abundant.

WELSH ONION/ JAPANESE BUNCHING ONION

(*Allium fistulosum*)
Partial sun. Cool. Average

A perennial alternative to spring onions (scallions) that will continue to split and divide creating more to harvest. A striking red-skinned option is available.

—

DWARF PEAS

(*Pisum sativum*)
Partial sun. Temperate. Average

Providing the same delicious crop as their larger pea relatives but on a far more manageable scale, these plants do not need staking.

Herbs

ROSEMARY

(*Salvia rosmarinus*)
Full or partial sun. Temperate. Average to dry

Available in two growth habits, the classic upright shape and also a lateral snaking prostrate form which is architectural and great for smaller spaces.

—

THYME

(*Thymus vulgaris*)
Partial sun. Cool. Average

Both classic and lemon-scented varieties are available in the standard clump-forming growth habit or creeping, matt-forming habit.

BASIL

(*Ocimum basilicum*)
Full or partial sun. Warm. Average

Go classic with Italian green or purple varieties or more adventurous with exotic-scented options such as cinnamon, lemon, lime or aniseedy Thai basil. For larger spaces try giant 'Lettuce Leaf'.

—

OREGANO

(*Origanum vulgare*)
Full or partial sun. Warm. Average

The classic pizza herb, great for drying to use later down the line.

—

CHIVES

(*Allium schoenoprasum*)
Partial sun. Cool. Average

A versatile way to add a gently oniony flair to any dish, the pink pom-pom flowers are also edible.

—

SAGE

(*Salvia officinalis*)
Full or partial sun. Warm. Average

Decorative and packed with flavour, sage comes in green, gold, purple, variegated, plus a white, green and pink-striped variety called 'Tricolor'.

—

MINT

(*Mentha species*)
Partial sun. Cool. Average

Has a surprising array of flavours from ginger and pineapple to apple and grapefruit, but bear in mind that this plant can be a thug and if growth is unchecked can verge on invasive. Grow in a container to keep it under control.

CHAMOMILE LAWN
(*Chamaemelum nobile*)
Full or partial sun.
Temperate. Average

A non-flowering, creeping variety of chamomile that is highly aromatic and just as useful as its full-sized cousin.

Spices

SICHUAN PEPPER
(*Zanthoxylum simulans*)
Full or partial sun.
Temperate. Average

Hardy shrub or small tree that can be grown outside in a temperate climate for its iconic fragrant pink peppercorns. Just beware these can have some large thorns.

—

CAROLINA ALLSPICE
(*Calycanthus floridus*)
Full or partial sun.
Temperate. Average

Hardy shrub providing an alternative to cinnamon that can be grown outdoors in temperate regions. Note that the bark of the woody stems is the only edible part of this plant.

—

GINGER
(*Zingiber officinale*)
Full or partial sun. Warm. Moist

Tropical plant that you can grow on from the rhizomes bought in the market.

CARDAMOM
(*Elettaria cardamomum*)
Full or partial sun. Warm. Moist

Beautiful leafy plant that produces seed pods for drying that are traditionally used to flavour Southern Asian dishes.

—

CUMIN
(*Cuminum cyminum*)
Full or partial sun. Warm. Average

Pretty frilly white flowerheads followed by strong aromatic seeds that, once dried, can be used whole or ground.

—

TURMERIC
(*Curcuma longa*)
Full or partial sun. Warm. Moist

Aromatic rhizomes that can be eaten raw or dried and ground to powder.

—

LEMONGRASS
(*Cymbopogon citratus*)
Full or partial sun. Warm. Average

The leafy tops of this aromatic grass can be steeped in hot water to make a refreshing tea. The stalks of the grass can be cut away from the base of the clump to be used in cooking.

—

SAFFRON CROCUS
(*Crocus sativus*)
Full or partial sun.
Temperate. Average

A beautiful autumn-flowering crocus that grows well outdoors in temperate climates. Each delicate flower produces just three long pollen-rich stamens which are harvested to make the world's most valuable spice: saffron.

Edible flowers

PANSIES AND VIOLAS
(*Viola tricolor*)
Full or partial sun.
Temperate. Average

Flowers are edible whole and available in a rainbow palette.

—

NASTURTIUMS
(*Tropaeolum*)
Full or partial sun.
Temperate. Average

Flowers, leaves and seeds are all edible and have a peppery flavour.

—

CHRYSANTHEMUM
(*Chrysanthemum*)
Full or partial sun.
Temperate. Average

Stunning flowers in a huge range of shapes and colours: remove and eat just the petals.

—

MARIGOLDS
(*Tagetes*)
Full or partial sun.
Temperate. Average

Petals and leaves are edible either raw or cooked and can add colour to dishes in a similar way to saffron.

—

DWARF SUNFLOWER
(*Helianthus*)
Full or partial sun.
Temperate. Average

Offering edible petals followed by large quantities of tasty seeds. Remove spent flowerheads to encourage more to be produced.

FUCHSIA
(*Fuchsia sp*)
Full or partial sun.
Temperate. Average

Ornate hanging flowers that are best enjoyed by just eating the petals. Flowers are followed by dark purple berries with a fragrant strawberry-like flavour.

Edible houseplants

PURPLE SHAMROCK
(*Oxalis triangularis*)
Partial sun. Temperate. Average

This is a common and particularly striking houseplant forming a clump of deep purple clover-like leaves. Though this wood sorrel is a regular on the 'plant selfie', it is less commonly considered an edible. Edible, it very much is, with leaves that taste like sour cherry or apple making an amazing addition to sweet and savoury dishes.

—

ALOE VERA
(*Aloe barbadensis miller*)
Full or partial sun. Warm. Dry

Spiky succulent with seriously soothing effects for both outside and inside the body. Can be prone to mealy bug infestations so keep an eye on it to catch it early.

—

AGAVE
(*Agave*)
Full or partial sun. Warm. Dry

Sculptural succulents that can be used to make a sweet syrupy sugar alternative.

TARO
(*Colocasia esculenta*)
Full or partial sun. Warm. Moist

Elephant ears is a much-loved houseplant with an edible corm, just make sure it is a colocasia not an alocasia as though similar in appearance, only the former is edible.

—

CANNA LILY
(*Canna indica*)
Full or partial sun. Warm. Moist

Though called a lily, Canna is grown more for its foliage than its flowers, and comes in a wide selection of eye-popping shades. Plants spread quickly, making it easy to justify harvesting some of the delicious rhizomes each season.

—

VANILLA ORCHID
(*Vanilla planifolia*)
Full or partial sun. Warm. Average

Climbing orchid native to Mexico and Central America offering delicate yellow or white blooms followed by intensely flavoured seed pods. Needs support.

COFFEE
(*Coffea arabica*)
Full or partial sun. Warm. Average

Glossy deep-green foliage plant, when flowering the stems become a profusion of stunning white flowers that are followed by bright red berries containing a hard bean that can be roasted and ground to make your classic morning pick-me-up.

Clockwise from top:
Chard; Marigold;
Taro; Cardamom;
Saffron.

PROJECTS

Cloning Store-bought Herbs

Don't panic, there's no need to crack out the lab coat or embrace your inner mad scientist, *cloning* is not as sci-fi as it sounds, though the sciency bit is interesting so I will indulge you with it. Cloning is a super-easy and cheap way to multiply your herb collection pretty much infinitely. The reason I refer to this process as cloning is because it's genetics. If you were propagating new plants from saved seed, the resulting seedlings would combine inherited traits from two parents to display a unique set of characteristics. But the process I am describing here creates a brand-new plant from a small section (or cutting) of the parent plant, and so any new plant will be genetically identical to the parent, and therefore a clone. If you want to get your official horticulture pants on, this is called vegetative propagation, and it works for a wide range of essential herbs. Right, that's the science! Supermarket herb plants are cheap and widely available and the benefit of using this technique is that you can make new plants until you run out of space. However, to avoid disappointment, this won't work for parsley, chives, dill or coriander (cilantro). On the other hand, you are almost guaranteed quick success from basil and mint – they love to be propagated in this way, and I recommend starting your cloning journey with one of these to get the hang of it. Here, I've used basil. A little trickier to clone are woody herbs such as rosemary, oregano, thyme and sage: these will root but you should expect slower and more mixed results.

Difficulty Level
Easy

You Will Need
The Tools
Snips or scissors
Glass jar or small vase

Suitable Plants
Basil
Mint
Rosemary
Oregano
Thyme
Sage

STEP 2
Take cuttings from the longer stems of your parent plant.

STEP 3
Snip off all the leaves from the bottom two thirds of your cuttings.

STEP 4
Place your cuttings in water.

STEP 6
Now it's a waiting game for the roots to grow.

MAINTENANCE

If you water your army of little clones regularly and feed it every two weeks with a good organic liquid feed, it will repay you with quick, strong growth. You will be harvesting lush handfuls of fragrant herbs in no time. Regular haircuts will encourage your plant babies to branch out and become bigger, bushier beasties.

The immortal tomato

You can use this technique to keep your favourite tomato variety going indefinitely! As tomato plants grow, they send off little side shoots at each node, where each leaf attaches to the main stem. When a side shoot reaches 15 cm (6 in) long, snip it off where it attaches to the parent plant, and treat it as if you were cloning a herb. Within a week, substantial new roots will emerge. As one plant nears the end of its life, you can use a cutting to start a new plant.

HOW TO DO IT

1 Store-bought herb plants are not grown to last. Their pots usually contain too many plants sown too closely together and have been forced to grow in a simulated environment in a rush to reach the store's shelves. Time, after all, is money. Sadly, this doesn't make for strong plants. So, the first thing to do is to give your parent plants some love: repot them into a larger container so they have room to stretch their legs; and give them an organic liquid feed so they start to grow more strongly.

2 Using clean snips or a pair of scissors, take cuttings from some of the longer stems of your parent basil plant. Taking several cuttings will increase your chance of success. You want to make your cut just below the node on a stem. The node is the place where the leaves attach to the stem. Ready for another sciency bit? The areas below a node have higher concentrations of auxin which is basically a plant growth hormone. These growth zones can be encouraged to change their nature and start growing roots instead of leaves. These roots are called adventitious roots, that's basically a root that grows from an area of the plant which doesn't usually produce them. Told you, pretty interesting stuff!

3 Carefully snip off all the leaves from the bottom two thirds of your cuttings. Additionally, if your cutting has any big leaves, a bud or flowers at the top, snip these off too, leaving only the smaller leaves. You want to redirect as much of the cutting's energy as possible into growing new roots, so you don't want it to flower or set seed at the same time. As a bonus, add the leaves you removed to a delicious dish and you can immediately enjoy the fruits of your labour.

4 Next, place your cuttings in water, making sure that at least one of the nodes is under the water level but that none of the leaves is submerged. Submerged leaves can start to rot and this can kill your cutting.

5 Place your cuttings in a warm, light spot but out of direct sunlight.

6 Now begins the waiting game: how long you will have to wait depends on the herb in question and environmental factors such as temperature. Basil and mint can start to form roots in a week or less, rosemary cuttings can take an agonising eight weeks. Don't worry if it looks like nothing is happening for ages, as long as the cutting is still alive at the top then your clone babies are working away at growing new roots. Change the water every few days to keep it fresh and clear of algae.

7 Once the roots are about 3 cm (1 in) long it's time to pot them up. To do this, fill a small pot with peat-free compost. Make a hole in the compost in the middle of the pot using a dibber, pen, pencil or finger. Being careful not to damage the delicate roots, place your rooted cutting into the hole so that all its roots are inside. Gently turn the cutting as you lower it into the hole to help the roots wrap around the stem and go into the hole without damage. Cover the roots with more compost and lightly firm it down so your cutting doesn't wobble.

8 Water and place back in a warm, light position. Congratulations, you have made a brand-new, little plant. Repeat as many times as you wish – it can get quite addictive.

Self-watering Pots

The longer you read this book the more you will see the magic word, *watering*. The more plants you share your life with, the more you will engage in this activity. Honestly, it's not always easy to stay on top of watering duties, in this fast-paced, time-poor society we live in, but don't worry you busy little bees, I've got you covered. It is really easy to make a pot that will do 99 per cent of the hard work for you. You can transplant any existing herb plants, edible flowers and small houseplants into these self-watering containers. Additionally, they are the perfect planter for any freshly rooted herb babies in the cloning project.

Difficulty Level
Easy

You Will Need
- Wide-necked glass vase with a 10 cm (4 in) opening
- 12 cm (5 in) terracotta pot with drainage hole
- Rope or thick cord made of natural fibre such as cotton, hemp or bamboo
- Peat-free multipurpose compost
- Slow-release granular plant feed
- Teaspoon, dibber or pencil

Suitable Plants
Herb cuttings (see page 47)

The Tools
Scissors

Fresh take

If you like to see what's going on beneath the surface, you can repeat this project but in place of flowerpots use basket planters. Basket planters are designed for pond plants and are essentially plastic pots with slits around the sides and base for water to permeate. They are sold in many sizes and shapes so you should be able to find one to fit snuggly into your wide-necked glass vase. Plant up your cutting as before (leaving out the rope) and place into the neck of the vase. As there's no rope, you want the bottom of the planter to always be touching the surface of the water, and to make sure the vase is always topped up with fresh water. As the cutting settles in you will see roots appearing at the basket's slits, then reaching down to the water and making a fascinating display.

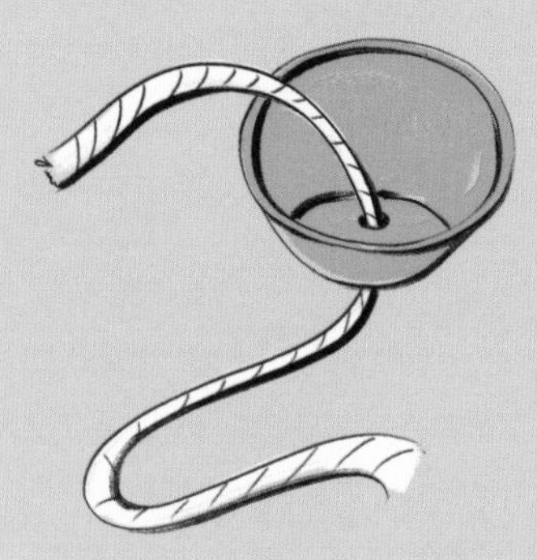

STEP 1
Feed the rope through the drainage hole in the bottom of the terracotta pot.

STEP 2
Coil the top half of the rope so that it spirals into the base of the flowerpot and up the sides.

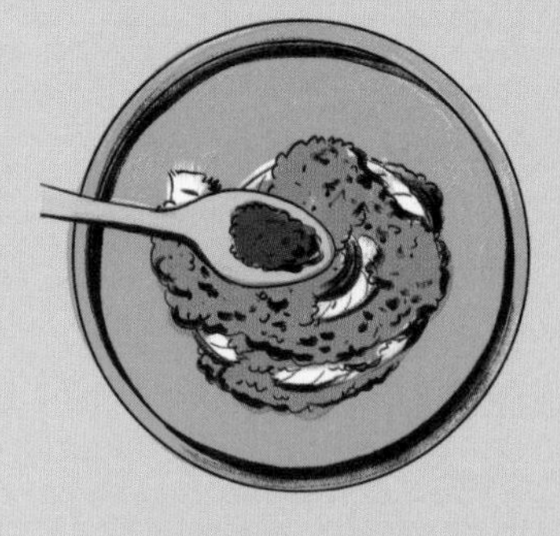

STEPS 3–4
Cover the coil with a handful of compost and sprinkle one level teaspoon of slow-release granular plant food across the top.

STEP 6
Fill the vase two-thirds full of water and feed the rope from the pot into the water.

STEP 7
Make a hole down to the base and insert your cutting without damaging its roots.

STEP 8
Firm the compost and top up with a little more if necessary.

HOW TO DO IT

1 Take the rope and cut a 120 cm (48 in) length. Feed the rope through the drainage hole in the bottom of the terracotta pot. The base of the pot should be at the halfway point of the rope, so you have 60 cm (24 in) sticking out of the bottom of the pot.

2 Firmly hold the base of the pot and the rope protruding from the bottom. Coil the top half of the rope so that it spirals into the base of the flowerpot and up the sides. If you find it tricky to get the rope to stay in place try twisting the rope as you coil it, this will help create a tight spiral.

3 Take a handful of compost and press it into the base of the pot, covering the coil. This will also help to hold the rope in place.

4 Sprinkle one level teaspoon of slow-release granular plant food across the top of the compost in the base. You won't be watering these plants, which means you won't be feeding them. This plant food will supply your plant babies with a steady source of vital nutrients over a long period. There are many options, but I recommend an organic-based feed if the end product is something to eat. I use seaweed flakes or – grizzly as it may sound – blood, fish and bone. This is a fantastic slow-release plant feed that your plants will adore. I know, non-vegetarian veggies, strange but true!

5 Fill the rest of your pot with compost.

6 Fill the vase two-thirds full of water and carefully feed the bottom half of the rope from the pot into the water until the pot is sitting within the vase's rim. Make sure the bottom of the pot is not touching the surface of the water. You may find that the rope floats to begin with, and it can take some time for the rope to saturate and begin to draw the water up and into the pot. This is normal, don't worry.

7 Using a dibber or pencil, make a hole in the middle of the compost all the way down to the base of the pot, circling the dibber or pencil to make the hole broad enough to insert your cutting without damaging its roots. Lower your rooted herb cutting into the hole. If this is tricky, try twisting the cutting as you lower it so that the roots spiral around the stronger stem.

8 Firm the compost around your cutting and top up with a little more if necessary. Give the cutting a little water, not a drenching, just to keep the cutting moist while the rope is becoming saturated and starts to wick the water up into the pot. This is the only time you will need to do this.

9 Place the pot in a bright spot but away from direct sunlight as it can cause algae to grow in the reservoir. Although algae won't affect the plant's growth, it can become unsightly and permanently stain the rope wick.

MAINTENANCE

Top up the reservoir in the base of the vase as this supplies the plant's food and water. It is best to do this before the water level drops too low. If it dries out entirely you will need to wash out any unsightly residue in the base of the vase and re-saturate the rope. If the water turns green or you see a residue on the sides of the glass, this can be the start of algae. Give the vase a good wash and fill with fresh water. Give your herb cuttings regular haircuts to keep them tender and encourage them to become bushy and productive.

Salad Trolley

Difficulty Level
Easy

You Will Need
Trolley
Pots and containers
Expanded clay pebbles
Peat-free multipurpose compost
Metal S hook

The Tools
Silicone bathroom sealant

Suitable Plants
Lettuce
Rocket/arugula
Basil
Nasturtium
Pansy
Chrysanthemum
Dwarf sunflower
Celery
Tumbling tomatoes
Fuchsia

The humble salad has been humble too long! Salads can be so much more than lettuce alone and are as much a treat for the eyes as the mouth. Edible flowers offer diverse flavour profiles and are a thing of beauty while growing and on the plate. Many also offer a second crop to enhance salads: after their bright, beautiful flowers die back, the roasted seeds of sunflowers are a fantastic, nutrient-rich, crunchy addition to the salad bowl. Fuchsia flowers put on a stunning display and are followed by an abundance of fragrant berries to give a zesty twist to salads. Here I have designed a petite and colourful, one-stop shop for the sexiest salads you can eat. As this little salad garden is on wheels, it can be moved around your home to make the most of natural light throughout the day while never getting under your feet.

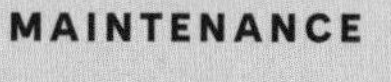

MAINTENANCE

Watering is necessary to keep your plants in tip-top condition. Remove dead flowers regularly to keep your pansies and chrysanthemums productive, stimulating more flowers over a longer season. The tomatoes and cucumber will benefit from a weekly feed once they start to flower.

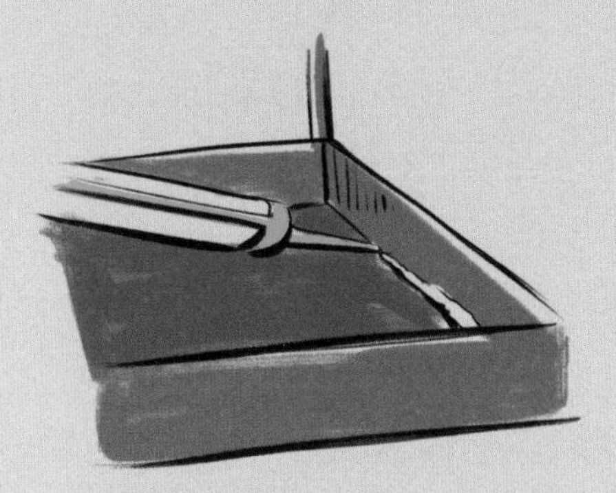

STEP 2
Use silicone bathroom sealant to seal around the base.

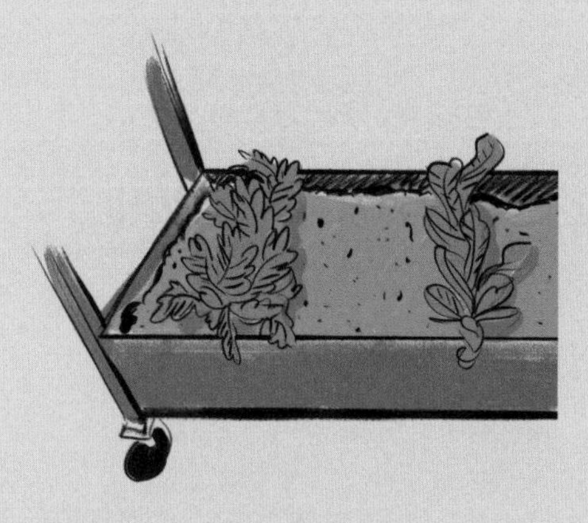

STEPS 3–6
Fill the middle shelf with cascading plants, the top shelf with taller crops and fill the base with compost to make a miniature raised bed. Plant up lettuce and rocket (arugula).

STEP 7
Fix an S hook onto the trolley to hang a handy pair of scissors for instant harvests.

HOW TO DO IT

1 Build or assemble your trolley following the manufacturers' instructions.

2 Using the silicone bathroom sealant, seal all around the base and sides on the inside of the bottom shelf of the trolley, so that it is watertight. Allow the sealant to dry.

3 Fill the middle shelf with cascading plants. Use pots of nasturtiums and fuchsias to tumble over the ends and tumbling tomatoes along the longer sides.

4 The top shelf is the spot for taller crops such as an eye-catching dwarf sunflower (ensuring it's a dwarf variety). Surround your sunflower with basil, celery and chrysanthemums which offer edible flowers and leaves to give dishes a citrusy zing. In the spaces left on the top two shelves, squeeze in pots of bright pansies.

5 Once the silicone sealant is dry, cover the base of the bottom shelf with expanded clay pebbles to a depth of around 2 cm (¾ in) to form a drainage layer.

6 Fill the rest of the base with peat-free multipurpose compost to make a miniature raised bed. Plant up pretty rows of lettuce and rocket (arugula). You can harvest from these over a long season if you just take the outer leaves as you need them.

7 Fix an S hook onto the trolley to hang a handy pair of scissors for instant harvests. Enjoy glorious salads that evolve in taste and beauty through the season, along with your plants.

Edible flowers: much more than a pretty face

Sunflower	Unopened buds can be cooked and eaten and have a flavour similar to globe artichokes. Subtly nutty-tasting petals are followed by flowerheads laden with seeds that are fabulously tasty if roasted.
Lilac	Highly-perfumed, little blossoms give a fragrant flourish to desserts.
Camellia	Delicious used fresh as a garnish and dried in Eastern cuisine.
Roses	With their classic, Turkish-Delight flavour, rose petals are a traditional addition to Middle-Eastern desserts. The hips that follow the flowers can be used in preserves.
Fuchsia	Best enjoyed by removing all the green parts, these stunning flowers are followed by fragrant, dark-purple berries.
Pansy	Pretty as a picture with a lettuce-like flavour, the more flowers you pick, the more you get.
Viola	Masses of pretty little flowers that taste like lettuce and are great in salads.
Marigold	These citrusy flowers and leaves are the perfect addition to salads.
Calendula	Will add both colour and a peppery flavour to cooking. The petals can be dried and used as a substitute to saffron.
Citrus	Overwhelmingly scented, these architectural flowers provide both decoration and flavour to cakes.
Tuber rose	One of the most perfumed flowers in the world. These are highly prized by perfumers and a traditional seasoning in Indonesian cooking, particularly as a flavouring for soy sauce.
Borage	These cucumber-flavoured flowers are fantastic in drinks and salads.
Daylily	The ultimate edible flower: its unopened buds are crisp and lettucey; in full bloom the flowers are added to hot and cold dishes; even the withered flowers are dried to use as seasoning.
Lavender	A fabulous way to add floral notes to baking and desserts.
Nasturtium	Both flowers and leaves are edible and give a strong, peppery hit, similar to rocket (arugula) or watercress.

Microgreens

Ok, I admit it, I'm not always a patient person! I can get seriously antsy waiting for seedlings to grow, mature, flower and ultimately set fruit, all the time itching for that inevitable moment when I enjoy the fruits of my labour. I love every agonising moment of it. However, some of my harvests can take years to arrive, so it is great to have a quick harvest ready for when anticipation gets too strong. Microgreens are the perfect fix, providing a bountiful harvest in as little as a week from sowing. Plus, they are considered a superfood! Microgreens generally contain a higher density of vitamins than the same weight of mature vegetables, so they are a great way to add a nutritious punch to your meals. You can grow these tiny crops in many ways, but these are my go-to methods, one with growing media, one without.

Difficulty Level
Easy

You Will Need
Containers without drainage holes
Coconut coir (for method one)
Hemp mats (for method two)
Shallow tray
Grow lights (for method two)

The Tools
Mister or spray bottle
Self-adhesive window shelf

Suitable Plants
Seed of any plant with edible leaves (see page 66)

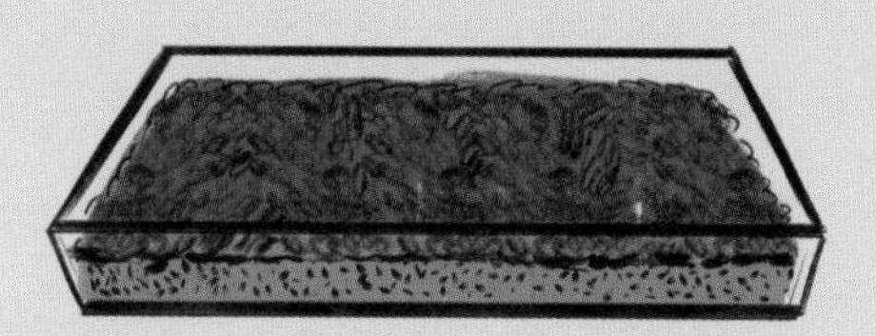

STEP 2
Take your selected container, fill it two thirds full of expanded coir and gently press it down.

STEP 3
Sow your seeds. You can use any seed of which you eat the leafy part of the plant.

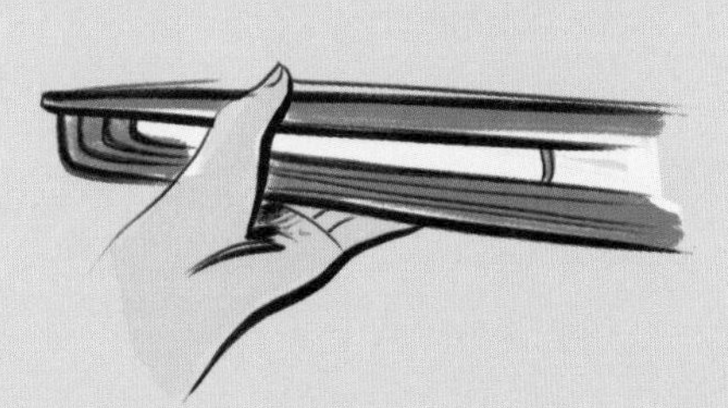

STEP 5
Place your microgreens in a warm, light place. Close to a window is the perfect position.

STEP 6
Give your containers a misting every day to keep the moisture levels constant.

METHOD ONE: COCONUT COIR

1 Coconut coir is usually sold in dehydrated compressed blocks. Place the block in a deep dish or bowl and gradually add water until the coir is hydrated and fully expanded.

2 You can grow microgreens in almost any container, from seed tray to jam (jelly) jar to yoghurt pot, with the proviso that the pot must not have drainage holes in the bottom. Take your selected container, fill it two thirds full of expanded coir and gently press it down.

3 Time to sow your seed. The world really is your oyster: you can use any seed of which you eat the leafy part of the plant. This means all herbs, any leafy greens, root vegetables, peas, squashes, wheatgrass, even (my personal favourite) sunflowers. Mustard and cress are the fastest, easiest and most foolproof crops that can be harvested in less than a week, and these are great to start with. Sprinkle your selected seed evenly across the surface of the coir, don't be shy, get plenty of seed in there.

4 Many microgreen seeds can be surface-sown, however, if you are growing a large seed species such as peas, squash or sunflowers cover with an additional thin layer of coir, just enough to cover the seeds.

5 Place your microgreens in a warm, light place. Close to a window is the perfect position but these tend to get crowded, so how to harness the light? We are going vertical again! I grow my greens stuck right onto the window (see page 150 for the Windowgarden suppliers). If this doesn't appeal, check out shelves for the bathroom: you can repurpose self-adhesive shelves for the shower to create gorgeous window gardens. When attaching the shelves to the window, clean both the glass and the suction cups with alcohol wipes to remove all traces of grease and dust and to ensure a strong grip: you don't want your greens to nosedive! The more greens you grow, the more shelves you can add to your window garden.

6 Now it's a waiting game – the time your seed takes to germinate depends on the type of seed sown. Make sure you give your containers a misting every day to keep the moisture levels constant and before long that rich brown coir will turn into a lush forest of tasty, vitamin-packed plant babies.

MAINTENANCE

Whichever method you use, harvest your greens when they are around 5 cm (2 in) tall, or 10 cm (4 in) tall for pea, sunflower or squash shoots. Use a pair of scissors to snip the greens at about 1 cm (½ in) above the roots. Add them directly to cooking and dishes for instant gratification. Once you have snipped all your greens you will need to sow a new crop of seeds, except for pea seeds which can be harvested twice before re-sowing.

METHOD TWO: HEMP MATTING

1 Hemp matting is the perfect medium to grow microgreens without soil. Its fluffy, fibrous top provides an ideal surface through which roots can twine, and it retains a good reservoir of water. The thin mats can be placed on any level surface and are perfect under grow lights, where no natural light is available, as illustrated in the farm desk project (see page 109).

2 Place the hemp mat on a shallow tray and spray with water until saturated.

3 Sprinkle your selected seeds on top.

4 Place the tray under full spectrum grow lights and mist with water daily.

5 Before long you will have a tasty and nutritious crop of microgreens.

Awesome microgreens to try

Cress	The fastest and easiest microgreen and a winner in an egg sandwich. Harvest in 5–7 days.
Radish	Super-easy, spicy and pretty too, in pale greens through to bright pinks. Harvest in 7–10 days.
Amaranth	Most visually striking to grow, choose one of the red varieties for nearly neon sprouts. Harvest in 10–14 days.
Beetroot (beet)	Beautiful, red-stemmed or fully red leaves with a deep earthy flavour. Soak seed overnight before sowing. Harvest in 10–14 days.
Chard	Incredible rainbow-stemmed leaves with a soft earthy flavour. Pre-soak seed overnight. Harvest in 10–14 days.
Pea shoots	Delicious soft pea flavour. Allow to regrow after the first harvest as pea shoots crop twice. Grow in coir rather than hemp mats. Harvest in 14 days.
Squash	Crisp and tasty in cooking or eaten raw. Grow in coir rather than hemp mats. Harvest in 14 days.
Sunflower	Delicious, crisp, nutty and my favourite. Harvest before the first true leaves appear. Better for coir than hemp mats. Harvest in 14 days.
Kale	Everyone's favourite supergreen, nutrient-packed in miniature. Harvest in 10 days.
Rocket (arugula)	Adds a peppery punch to food before this crop can bolt. Harvest in 7 days.
Spinach	Mild but nutrient-packed leaves. Harvest in 10 days.
Fennel	All the aniseed flavour of the mature plant but in less time, space and effort. Harvest in 10–14 days.
Basil	Intense flavour in tiny green leaves and for fancier sprouts try a purple variety. Harvest in 10 days.
Coriander (cilantro)	A serious flavour punch to sprinkle on hot dishes. Harvest in 14 days.

Indoor Spice Garden

Spices are the epitome of flavour. They are intense, aromatic, invigorating and for thousands of years, valued as highly as gold. Why? Well, even to this day, most spice plants grow almost exclusively in tropical regions. It may be easier to ship these products around the world today, but as we become more aware of our food's carbon footprint, the question is less about the impact on pocket and more about the cost to our planet. Any reduction in our food miles is a step in the right direction. The good news is that it's easy to grow a diverse spice garden in the warmth of your home. Even better news, many of these plants can be propagated cheaply from seed or spices found in the supermarket. Spice plants are beautiful in their own right and will make stunning houseplants to literally spice up your life!

Difficulty Level
Intermediate

You Will Need
Pots and planters

Suitable Plants
Ginger
Cumin
Carolina allspice
Lemongrass
Cardamom
Saffron
Sichuan pepper
Turmeric

HOW TO DO IT

1 Ginger – The fresh ginger root you use in cooking is actually a dormant rhizome of the plant, but with a little love and patience it can be coaxed to produce new plants. When selecting your ginger root (though it's not actually a root), aim for an organically-grown piece, and this may be easier to find in a farmers' market or health food store, though some of my most productive ginger was found in Asian supermarkets. Select the plumpest, healthiest roots, because the drier and more shrivelled they are, the less likely they are to sprout. You also want the knobbliest pieces you can find; as each knobble is, in fact, a node from which the shoots will sprout. Place your ginger in a warm, bright, sunny spot and wait. In as little as two weeks, green shoots can start to sprout from the nodes, though it can take much longer, so be patient. If the rhizome begins to turn green you know that shoots are on their way. When you finally have shoots you can plant the rhizome shallowly in a pot. Rhizomes grow just under the surface of the soil so fill a pot with compost, place the ginger on top, and add just enough compost to almost cover it. Ginger grows best in the Tropics so place it in a warm, humid environment, watering well and keeping the compost moist. Once ginger is established you can harvest it at any time, but the longer you leave it, the more you will have, so try to harvest small pieces and replant the rest.

2 Lemongrass – You can buy new lemongrass plants in spring, when they are generally available, but they are easily started from seed. Simply sprinkle seed sparingly onto the surface of some moist compost in a small pot. Place the pot somewhere bright and warm and keep the compost moist – before long new seedlings will appear. It is also possible to propagate lemongrass from sticks bought in the supermarket. Aim for an organic source if possible, remove all loose outer leaves, and submerge the sticks in 5 cm (2 in) of water. Change the water every few days to keep it fresh and free from green algae, and before long little roots should start to grow from the bases of the sticks. Once these are around 2–3 cm (¾–1 in) long, simply plant your lemongrass stalks in pots of compost, taking care not to snap their precious roots.

3 Cardamom – It has recently become possible to buy cardamom plants. The spice is the seed, so harvest whole seed pods once they start to turn brown and leave to dry fully.

4 Cumin – This is an annual flowering plant that grows between 15–60 cm (6–24 in) in height. It grows from seed, so you need to sow a fresh batch each year. Sow seeds in moist compost, around 1 cm (½ in) deep, keeping them well watered in a warm, bright spot. As your cumin babies grow, pot up as necessary. After flowering, harvest when the seed pods turn brown. Leave the pods to dry fully for several weeks before revealing the aromatic seed inside.

5 Cinnamon – Ok so no, you can't actually grow cinnamon at home, it comes from the bark of several species of the *Cinnamomum* genus of trees which reach 10–15 m (33–49 ft) in height – making them tricky house guests. However, a great alternative to cinnamon is Carolina allspice (*Calycanthus floridus*), a beautiful and easy to grow shrub. Though they have an intoxicating, cinnamony smell, the leaves and the purple-black flowers and the fruit are all poisonous. It's only the woody stems which can be cut and dried and, the bark then crushed and used as the perfect cinnamon stand in.

6 **Saffron** – Saffron is worth more than its weight in gold, but it is simple to grow at home. This valuable spice is harvested from the stamens of the saffron crocus but each flower produces just three stamens a year. Saffron crocus bulbs are fairly easy to buy with purple or white flowers. Grow them in containers, planting the bulbs pointy end up, 10 cm (4 in) deep in compost. We usually associate crocus with springtime but saffron crocus are autumn-flowering. Once flowering has finished, the bulbs will die back to a dormant state – don't worry, this is normal, they will be back!

7 **Sichuan pepper** – This highly-prized, aromatic pepper has a numbing effect and is actually easy to grow. Many varieties of Sichuan pepper or winged ash are available (see suppliers' page 150) and many are highly ornamental. *Zanthoxylum piperitum* 'Purple Leaf' for example has such dark foliage it is almost black, while *Zanthoxylum simulans* provides the perfect pink peppercorn: each has a unique and exciting flavour. Harvest the peppercorns when the seed casing opens, revealing the black peppercorn inside, and leave to dry. It is the seed casing as opposed to the peppercorn that carries the Sichuan sensation so don't get rid of these. This wonderful spice can last many months if stored in a dry, airtight container.

8 **Turmeric** – This striking yellow/orange spice can be grown from store rhizomes in much the same way as ginger, and harvested in the same manner too. There is no need to wait for these orange finger sized roots to send out shoots, just shallow plant directly into compost and keep warm and moist. Shoots can take a while to appear so be patient, but when they do, you will be rewarded with a beautiful lush green foliage plant.

MAINTENANCE

Ginger, cumin, cardamom turmeric and lemongrass are all tropical plants, so keep them warm and constantly moist. Carolina allspice, saffron crocus and Sichuan pepper hail from more temperate regions so can tolerate lower temperatures and drier compost, though they will be happy in the same conditions as your other spices. Feed your spice garden once a month to help it thrive.

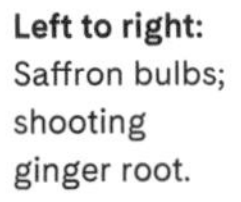

Left to right: Saffron bulbs; shooting ginger root.

What's in a name?

Well, quite a lot actually. All plants have a Latin name consisting of two parts (which is why it is called the binomial system). The first part indicates the genus and the second part is the species, and these names are usually italicised. Despite coming first, the genus name is a bit like a plant's family name or surname: there will be other plants with the same surname but they will have another name too so you can tell them apart, just like in a human family. This other name is the species name and, yes, confusingly comes second. Sometimes a plant will also have a cultivar name, which will appear in quotation marks after the genus and species. So basically, if I was a plant I would be *Hutchings Lucy* 'She Grows Veg'. Now, here's the thing, most people in this world do not walk into their local garden store and say, 'Excuse me, please can you sell me some *Solanum lycopersicum* plants? Oh, and if you have any *Helianthus annuus* seeds, that would be awesome.' No, you would go and ask where you could find the tomato plants and sunflower seeds. Most of the plants we are familiar with, and therefore likely to grow, also have a common name, such as marigold, ginger, lettuce and kiwi. But the horticultural industry loves using Latin names: it seems to be a badge of honour to converse in botanical Latin and common names are discouraged. I think this needs to change. Although the binomial system plays a vital role in ensuring we source the correct plant – there are, for example, a myriad ginger lilies but only a few are edible – the fact remains that botanical Latin can be intimidating and confusing to the newbie grower, and I hear this feedback from many of my followers. I am devoted to getting more people growing, so as a result you will always find me using a plant's common name where it is safe to do so. Where necessary, I have also included the plants' botanical Latin names so that you can source exactly the right thing. I hope this helps. Let's hear it for common names and making horticulture more accessible!

Hanging Window Herb Garden

We know that one of the essentials we need to supply plants with to help them thrive is light, it's a non-negotiable as far as plants are concerned. Windows are the perfect free source of life-giving light in our homes and so, inevitably we often find ourselves focusing our growing efforts around the windowsill. However, what do you do when you have reached that inevitable stage where your precious windowsill real estate is full to bursting but your plant collection is still ever growing? Well, here is a ridiculously fast and easy way to maximise all that space and light and turn your whole window into a growing area.

Use an over-door hanging storage rack to create a beautiful vertical window herb garden that also offers you total flexibility as it can be easily moved around the home or even from indoors to out as you choose. It can provide an equally space-saving solution outdoors as well as in the home, offering a way to utilise external window space, can be hung on a sunny wall or fence to form a green wall or can even add rows of growing room to your greenhouse.

Difficulty Level
Easy

You Will Need
Over-door hanging storage rack
16 x 12 cm (5 in) pots depending on the design of the rack
Peat-free multipurpose compost
Expanded clay pebbles (see page 150 for exactly what I used)

Suitable Plants
Any herbs

HOW TO DO IT

1 First select an over-door hanging storage rack. There is a wide selection of these available for use in kitchens and bathrooms, they generally cost very little and are available in a range of designs. I am using a classic and widely available design and probably the most common on the market (see page 150 for suppliers). Its simple wire mesh construction is attractive yet suitably strong enough to support a whole heap of plant babies. Over-door storage racks come with integrated hooks at the top and are designed to hang from the top of a door, adding rows and rows of storage space without obstructing the ability to open and close said door. They are therefore indeed perfect for use in exactly that way so if, like me, you have a glass door, you can suddenly make use of all that light and space you would never usually be able to utilise, whilst still retaining a fully functioning door. Alternatively, the hooks can just as easily be used to hang the rack from the top of a window frame or over the top of a sash window. Failing that, a simple wooden baton can be installed at the top of the window frame, to provide something to hang it from.

2 The next thing to figure out is how many pots your rack can hold. As a general rule, racks tend to fit three to four 12cm (5 in) pots on each shelf of the rack with a 4-shelf rack like the ones I use, that gives you enough for 16 plants. When selecting your pots there are two things to bear in mind. Firstly size, be sure they fit comfortably into the shelves; it is worth measuring this as it can be crazy frustrating to find that they are just slightly too big and therefore won't sit down fully into the available space. Secondly, if your rack is destined primarily for indoor use, use pots with no drainage holes on the bottom, this will ensure you avoid inadvertently creating a spectacular indoor waterfall when it comes time to water. For outdoors use, aim for pots with integrated drainage holes for exactly the opposite reason, the waterfall effect from one layer to the other will ensure all your plants get a good drink while saving you a bit of time and effort – see page 17 for more information on pots and drainage.

3 Now it's time to select your plants. You can grow pretty much any herb in here, many of our staple and most loved herbs hail from the Mediterranean so they will thrive in this window garden, basking in the glorious sunshine streaming in through the window. Clump-forming herbs such as thyme, Greek basil, oregano and marjoram make a particularly good choice as they will be easy to maintain in a neat clump that won't outgrow the available space between the shelves. If you are a rosemary fan then go for a prostrate variety, its sideways growth habit will mean it cascades beautifully out of the window garden and won't have to be trimmed as brutally as its upright cousin in order to keep it within its allotted space. For herbs with a more upright growth habit such as sage, chives, mint, basil, coriander (cilantro) or parsley, the top shelf or your window garden is the perfect spot to position these, they will be able to reach for the skies unobstructed. Want to position them a bit further down the garden? No problem, just make sure to give them a regular hair cut to encourage a bushier growth habit and maintain the shape you want.

4 Pot up your herb plants into your selected pots, using peat-free compost and clay pebbles for drainage. Full potting instructions on page 19.

MAINTENANCE

Make sure to keep your plants watered. Many herbs can be relatively drought tolerant, however prolonged or regular periods of dryness can trigger some of them to flower. Although herbs in flower are still completely edible, it can change their flavour which is not ideal. A regular chop is a good thing, as the stems of some herbs such as thyme, rosemary and sage can become woody as they mature. Regular cutting will stimulate them to keep putting out fresh tender green growth that is perfect for eating fresh or cooking with. Plus, it will help maintain a nice neat little plant that fits prettily and perfectly in your window garden.

Fresh take

Why not try making a basil garden that will let you harvest more than enough of this fragrant herb to make your own homegrown pesto? Grow a different variety on each shelf of the garden to create a beautiful show of colours and textures and add a complexity of flavour to your pesto unlike anything you will have sampled before. Basil can get quite tall, anywhere from 20–75cm (7–30 in) so make sure to put your tallest varieties on top and encourage the rest to grow forwards out of the window garden shelf so that they have room to head up. Some interesting varieties to consider:

Greek basil	A small basil that forms a tight clump of very small leaves, perfect for the window garden.
Purple basil	A number of different varieties are available, but all have stunning deep purple foliage.
Lettuce leaf or Genovese basil	Giant basil with huge leaves as big as your hand.
Cinnamon basil	A gorgeous sweet basil with a hint of cinnamon scent.
Holy or Thai basil	Iconic staple ingredient in Thai cooking, it has a unique liquorish flavour.

Rolling Room Divider

Difficulty Level
Intermediate

You Will Need
Sturdy clothes rail on wheels with a solid base
Four 20 cm (8 in) planters
Two strong canes
Paint (optional)
Cable ties
Twine
Peat-free multipurpose compost
Expanded clay pebbles

Suitable Plants
Climbing beans

The Tools
Scissors

Room dividers were invented by the Chinese sometime between the 4th and 7th centuries to divide larger spaces into distinct areas of usage. As well as marking out space, these dividers were highly ornamental, considered free-standing works of art and only afforded in the wealthiest households. Though they gradually became lighter and more portable after being adopted by the Japanese, their decorative tradition continued until the 15th century when they became the ultimate accessory for the fashionable interior. These days, the room divider falls predominantly into the decorative camp, but with more people living an open-plan life, the ability to subtly, visually apportion space can be effective. It can create a level of privacy, mask off unsightly storage areas or simply mark a line between two areas such as a seating and a dining area. Here is an easy way to make the ultimate botanical living screen that can be easily and conveniently moved around your home as required, while providing a heavy crop of beans potentially all year round.

STEP 2
Canes should be 6 cm (2 in) longer than the width of the rail so they stick out 3 cm (1 in) on either side.

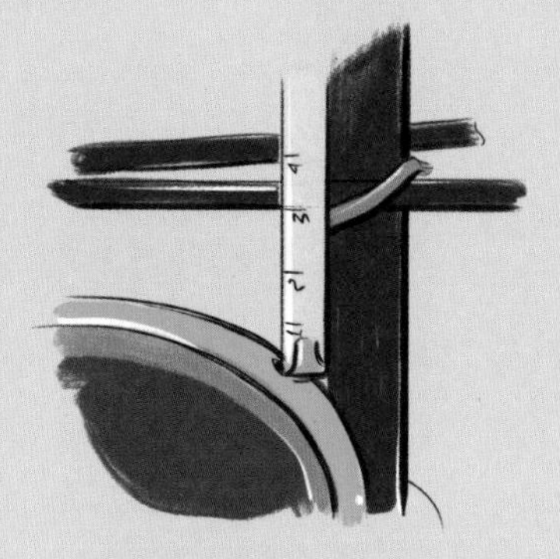

STEP 3
Secure the canes 7 cm (2¾ in) above the top of the planters with cable ties.

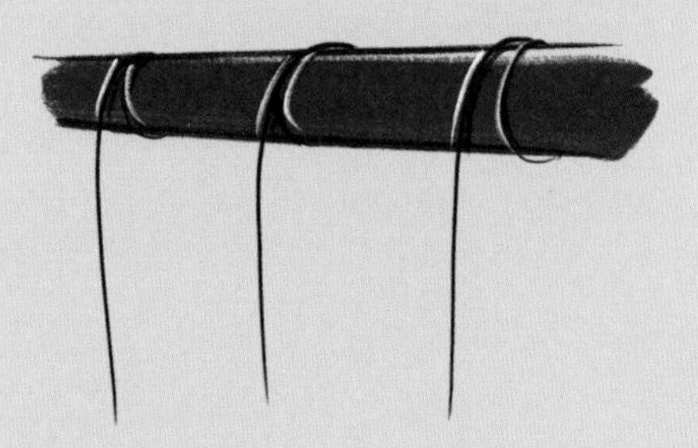

STEP 4
Wrap the twine around the cane.

MAINTENANCE

Food and water will obviously be required to keep your bean plants vigorous, and you will need to keep pinching out the tips as they get too tall for the rail. When the beans start cropping, make sure you pick them regularly. With a little assistance from a nearby grow light from mid-autumn through to mid-spring you should be able to grow fresh beans year-round. If any of the plants flag or stop producing beans, sow a new bean seed at the base and snip out the old plant. The new bean will soon take over where the last one left off.

Fresh take

Beans are not the only plant to try on your room divider – many climbing plants such as cucumbers or achocha will grow happily in containers. The really adventurous can try a small squash variety such as 'Sweet Dumpling'. For a low-maintenance option, perennial climbers such as passion fruit, kiwi and mashua can thrive if grown this way, or runner beans, yes, you heard me, runner beans are actually a tropical perennial and in the right conditions can live and crop for years.

HOW TO DO IT

1 When purchasing your clothes rail, ensure it has a base or shelf at the bottom to support the planters that will contain your vines. Also select a wheeled model so that you can move it easily. If your clothes rail arrives flat packed, and let's face it, it probably will, construct it following the manufacturers' instructions.

2 Measure your two canes against the width of your rail. You want your canes to be around 6 cm (2 in) longer than the total width of the rail so that they will stick out roughly 3 cm (1 in) on either side. If necessary, carefully cut the canes to length. At this point you can spray or paint your canes to blend in with your rail. I used a black rail and painted my canes black. This step is optional but can vastly improve the appearance of your screen.

3 Place the empty planters onto the base of your rail so you can see how high they reach. Place your canes on top of the planters, one on each side of the rail. Now use the cable ties to secure the canes 7 cm (2¾ in) above the top of the planters, fixing one on either side of the rail, pulling the cable ties tight. Remove the planters and place to one side for the moment.

4 Take the twine and tie it securely onto one of the canes at one end of the rail. Run the twine up to the top of the rail and wrap it all the way around the top bar before running it down to the cane on the other side. Again, wrap the twine once around the cane then it's back up to the top again and down the other side. Each time you go around the rail or a cane, do it about 5 cm (2 in) apart from the last time, in this way you will find you are creating vertical lines of twine all the way along the structure. Try not to pull the twine too taut, as this can cause the canes to bend upwards and will make your twine near the ends loose, just pull it tight enough so that it doesn't sag. For extra security I would recommend knotting the twine onto the canes at around every tenth turn. Continue in this way until you have vertical lines of twine running across the length of the rail and tie off the end of your twine onto one of the canes.

5 Fill your planters as per instructions on page 19 leaving around a 5 cm (2 in) gap at the top of the pot, and place them on the base of the rail in a row under the canes. Take your chosen bean seed and push one seed down into the compost: beans should be sown 5 cm (2 in) deep so that basically means pushing it down to the second knuckle of your finger. Do this under each length of twine on whichever cane is closest to you, ignoring the back cane. Cover the beans and water.

6 Cover the exposed soil with expanded clay pebbles to disguise the compost and deter soil-borne pests.

7 Your beans should germinate in around a week. Look out for telltale lumps on the surface of the pot to indicate your beans are about to pop out and say hello. Beans are fast growers, so before long will happily make their way up your room divider. Should any get lost along the way, simply wind them back in gently around the twine and they will soon spiral in the right direction.

8 When your beans reach the top of the structure and start grow above the rail, carefully pinch out the tip either with fingers or a pair of snips or scissors. This will encourage the plant to send out sideshoots lower down to fill out the rolling green wall.

Growing Food Over Arches

Difficulty Level
Intermediate

You Will Need
Strong garden arches
Strong cable ties

Suitable Plants
Small to medium pumpkin varieties
Gourds
Beans
Cucumbers
Melons

The Tools
Spade

Space is a precious commodity and something we all want to maximise. In garden design, space is divided into masses and voids. Masses are form: they are flowerbeds, raised beds, planting areas, trees, ponds and garden structures like sheds, sculptures and pagodas. Voids are the spaces in between: lawn, paths, patios, decking and the like. To create visual balance in a garden you need the right balance of masses and voids: on a basic level, this is generally considered to be one third masses to two thirds voids. However, if we want to maximise the space we have, aren't voids wasted space? Well, there is a way to use some void space while retaining passable areas to move through your garden. By creating arches and tunnels over pathways you can claim back growing space while keeping open highways. Growing vertically can also solve the challenge of growing space-hungry crops like pumpkins, gourds and melons in reduced spaces. These three sprawling plants usually require several square metres (yards) per plant but by growing these accomplished climbers vertically you can decrease their footprint to 30 square cm (1 square ft). This is one of the most beautiful ways to grow food, creating dreamy green tunnels to explore and forage within: they draw the eye through them and can be a clever way to highlight and frame focal points in your garden. Despite popular belief, summer squashes and courgettes (zucchini) do not climb so are not suitable for growing in this way.

HOW TO DO IT

1 Select a strong, sturdy, metal or wooden arch. Don't be tempted to buy a cheap, flimsy arch as it will probably collapse under little strain or the slightest breath of wind. Even thin vines, such as bean plants, can be surprisingly heavy when in full swing and there is nothing more disappointing than watching an arch collapse and the loss of its precious plants and crops.

2 Using a spade, dig a trench 30–40 cm (12–16 in) deep on each side of the path. Place your arch in the trench so that it is in the correct position and level on both sides. Now fill in the trenches and firm down so that the bottom of the arch is buried securely in the soil.

3 You can turn an arch into a tunnel by attaching multiple units together with strong cable ties. Longer tunnels are stronger, more stable and good for heavier crops like squash and melon. Make sure each arch is level and flush to the next so you can't easily see the gaps. Securely fasten them together with strong cable ties at the top of the arch and in at least one place on each side.

4 Plant your crops right up against each side of the arch or tunnel. Ignore the recommended spacing on the plant label or seed packet, we are rebel gardeners and growing in this way allows us to break the rules. Squash, melons and gourds can be planted at 40 cm (16 in) intervals along the sides. Cucumbers can be 20 cm (8 in) apart, and beans need just 10 cm (4 in) between each plant.

5 If any of your climbing plants head in the wrong direction, redirect them gently by training them in, and if they resist, loosely tie them in position with a piece of string or twine.

MAINTENANCE

Lack of water is a problem for productive crops and will prevent them from setting fruit and the young fruits from maturing. Squash, melons and gourds are hungry plants so will need a feed every week or two. Beans are self-sufficient as long as you water them, but you need to keep harvesting: if you allow the beans to ripen inside the pods, the bean plant thinks it has successfully set seed and it stops producing new beans. So, the more you pick the more you get. The same goes for cucumbers, keep on picking! Squash and gourds prefer to be left for as long as possible on the vines, harvest when the plants start to die back in autumn but before the first frosts. Store squashes somewhere cool and dry and they can last for months: I've kept 'Tromboncino' for more than 18 months! How can you tell your melon is ripe and ready to eat? Easy – use smell. There is no mistaking the heady, sweet aroma of a juicy, ripe melon.

The best winter squash and gourd varieties for growing along a tunnel

'Tromboncino'	A hugely productive multi-purpose, giant, butternut squash. Pick young and use like a courgette (zucchini) or leave to mature to more than 1 m (40 in) long and to last more than a year. This variety needs gravity to grow straight so is perfect for vertical growing.
'Sweet Dumpling'	A miniature squash around the size of a grapefruit that starts green and white striped but turns orange in storage. It is the perfect single serving squash, great stuffed and roasted.
'Golden Acorn'	A delicious, golden, deeply-ridged squash that, when roasted, makes a great lower-calorie alternative to mashed potato.
'Uchiki Kuri'	A bright orange, onion-shaped squash with a nutty, sweet flavour.
'Blue Kuri'	Beautiful greeny-blue mini-pumpkins, the perfect size to feed a family.
'Delicata'	The sweetest of all squashes, the high sugar content of this medium-sized squash means it caramelises beautifully when roasted.
Spaghetti squash	The flesh of this unusual squash grows as a knot of tightly matted strings and can literally be made into vegetable spaghetti.
'Achocha'	You only need one plant of this staggeringly productive gourd! The small, elegant and spiky fruit can be picked young for salads as an alternative to cucumber. Mature fruit can be hollowed out and stuffed like (bell) peppers.
'Luffa'	Grow your own scouring sponges! The interior of this unusual, heat-loving gourd dries into a hard, fibrous sponge that's great for cleaning plates and even people. Pick gourds when young and tender to eat like courgetti.

Window Plant Shelfie

The average windowsill is frustratingly small. Plants need light, so we naturally gravitate towards the windowsills of our homes where we can guarantee sufficient light. However, window ledges fill up fast, especially as we get more excited about the possibilities of growing. So, how to make more of the light flooding in through the windows of our homes? Simple, make more sills. By installing a series of floating shelves within the window frame we can vastly increase the number of plants we grow. The sun beaming in around our thriving plant collection looks stunning, and this is a clever way to increase privacy from the world outside!

Difficulty Level
Easy

You Will Need
Wooden planks
Wooden batons
Screws

Suitable Plants
Any sun-loving plants

The Tools
Measuring tape
Saw
Drill
Screw driver
Spirit level

STEP 2
Measure the inside width of your window at the place that you want each shelf.

STEP 4
Measure the depth of your window frame.

STEP 5
Screw the batons in two positions into the window frame to ensure they cannot spin.

STEP 8
Add shelves and plants.

HOW TO DO IT

1 Pick the window you want to pimp and decide how many shelves you can fit within it. I recommend spacing shelves at least of 40 cm (16 in) apart, to allow growing room and to let in light, and increase this to fit in taller plants.

2 To cut shelves to size, measure the inside width of your window at the height you want each shelf. (Please don't assume all the shelves will need to be the same width – I can't tell you how many window frames are not exactly rectangular, even in modern houses.)

3 For the planks, pine is easy to get hold of and can be painted to match the window frame for a modern look. If you prefer a rustic style, look for reclaimed planks: I used reclaimed scaffolding boards which were easy to source. Using the saw, carefully cut your planks to the correct length: if you are buying modern pine from a hardware store you may be able to get them cut to the correct sizes.

4 Now measure the depth of your window frame. This is the length you want to cut your batons to.

5 Hold one of the batons in position on one side of the window frame and using the drill carefully screw it into position in two places to ensure it cannot spin.

6 Now place the shelf on top with a spirit level resting on it and position it so that it is completely level and mark with a pencil underneath the shelf on the other side of the window frame. This mark will be the position for the top of the other baton. Attach it with two more screws. If you want to make your shelves look extra slick, paint the batons to blend in with the window frame before placing your shelf on top.

7 Repeat step 6 until you have installed all the shelves.

8 Now arrange your plants and containers on the shelves so they can catch some rays.

MAINTENANCE

The increased heat and light that your plants receive in their bright, new location can dry them out more quickly. Keep an eye on them and water when the surface of the soil dries out. A good feeding routine, especially while flowering and fruiting, will keep them lush and productive.

Fresh take

If your window frame is not deep enough for shelves, or if you simply want an alternative take on making the most of your window space, why not install rails instead of shelves? You can suspend hanging pots or cutlery pots from these rails.

Kokedama Citrus Orchard

Kokedama literally means 'moss ball' in Japanese, which is a perfect description. The practice of encasing the roots of plants in a nutrient-rich, moss-covered clay ball is a modern evolution of the ancient Japanese art of bonsai, where mature trees are kept at a miniature stature by precise pruning and restricting the roots. The trend for exposing root balls grew from the principle of wabi-sabi, an appreciation and acceptance of the transience and imperfection of nature. This inspired a practice called nearai, where bonsai plants were grown in extremely restrictive pots until the root balls were tightly bound and sculptural, with little or no soil. The plants were then displayed on flat pieces of ceramic or driftwood so that the roots could be fully admired. This practice has evolved into the less brutal kokedama, which elevates plants to the status of living sculptures. Because they have no need for a pot, it is popular to suspend kokedama in groups to form string gardens. If you can find them, small citrus specimens are a great choice for growing this way. Citrus are fabulous houseplants: they are beautifully architectural with glossy, dark green foliage that exude perfumed oils, and in spring their long-lasting flowers release one of the strongest and headiest scents. Kokedama is a truly stunning way to create a mini citrus grove in your home. What could be more rewarding that sitting down to admire your Japanese sculptural handiwork, while sipping a G&T garnished with a freshly plucked home-grown lime? Heaven!

Difficulty Level
Tricky but worth it

You Will Need
Bonsai compost
Natural clay
Slow-release plant food
Sphagnum moss
Clear, thin, nylon fishing wire
Plant stand or strong cord/ribbon
Ornamental cord or rope

The Tools
Potting tray or large mixing bowl
Watering can
Baking cup measure
Scissors

Suitable Plants
Calamondin
Orange
Lemon
Lime
Yuzu
Finger lime

Plants must be shorter than 60 cm (24 in) and pots no wider than 19 cm (8 in)

STEP 1
Take your citrus tree out of its pot and remove all of the soil from the rootball without damaging the roots.

STEP 2
Measure seven cups of bonsai compost, two cups of natural pottery clay and a heaped teaspoon of slow-release fertiliser.

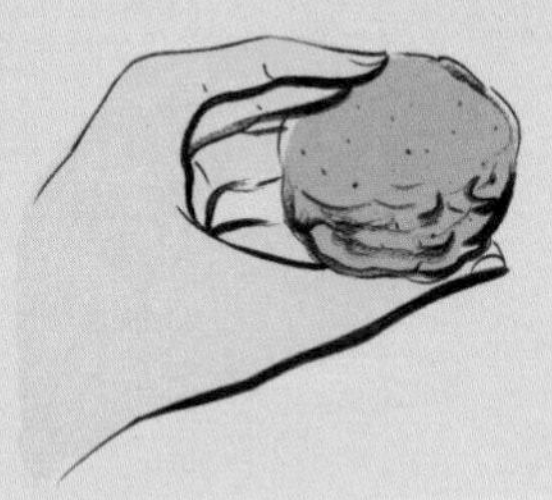

STEP 3
Knead your clay and compost and fertiliser together.

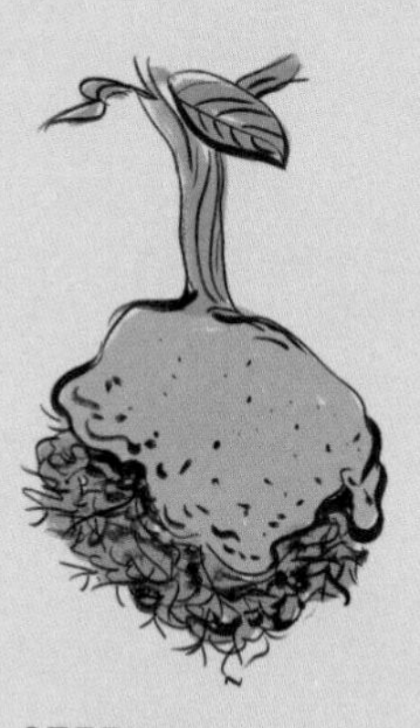

STEP 4
Use your clay compost mix to cover the roots.

STEPS 5–6
Moisten the clay ball and cover with sphagnum moss. Wind the fishing wire all round the ball to hold the moss in position.

STEP 7
Hang using a strong cord. Pass the cord down through the foliage and tie securely around the trunk just under the lowest branch.

HOW TO DO IT

1 Take your little citrus tree out of its pot and gently remove all of the soil from the rootball without damaging the roots. Try to gently rub and shake off all the loose soil from the roots' outer edges to make the next step easier.

2 Using a baking cup measure, put seven cups of bonsai compost onto a potting tray or large mixing bowl. Despite being tiny, bonsai are fully mature trees and need a full spectrum of nutrients in the small amount of soil available to them: bonsai compost supplies everything a mature tree needs and is especially water retentive. Now add two cups of natural pottery clay for kiln firing: this is easy to find online but should not be confused with air-drying clay which has polymers and chemicals added and won't work for kokedama. Finally, for extra nutrition, add a heaped teaspoon of slow-release fertiliser such as seaweed flakes.

3 If you made mud pies as a kid you are going to love this part! Roll your sleeves up and start kneading your clay and compost together, just like you would with cookie dough. Have a watering can on hand and add small amounts of water as you mix and knead your substrate. Aim for a well-combined, sticky texture that holds together firmly as a ball. If your mixture is still crumbly and won't hold together when both parts are combined, you may need to add a little more clay and water.

4 This is the tricky part, but stick with it. Take the plant and use your clay compost mix to slowly cover the roots. You don't want a lot of crumbly soil on the roots as it will prevent the clay mix from sticking and will keep falling off in chunks. Once you have completely covered the roots in the clay mix, form it into the most perfect ball shape possible.

5 Wet your hand and moisten the surface of the clay ball, to make it stickier. Don't pour water over it as your substrate will probably come sloughing off. Take the sphagnum moss and gradually cover the ball, bit by bit, pressing the moss into the moist, sticky clay.

6 Pass your fishing line once around the ball and tie a secure knot, don't cut the ends, one end should still be attached to your reel of wire and the loose end should be left around 15 cm (6 in) long. Now start to wind the fishing wire round the ball in all directions. This will hold the moss in position and stop the clay ball from disintegrating. You want all parts of the ball criss-crossed with supportive fishing line. When you are confident that you have circled the ball enough that it is secure, tie the line off to the loose end from the first knot and snip the ends.

7 To hang your kokedama ball, select a strong cord – here I use a black and gold nylon trim for dress-making. Fabric shops offer myriad trims and cords in every shape and style – a good opportunity to get creative with your string garden. Pass the cord down through the foliage of your citrus tree and tie it securely around the trunk just under the lowest branch. The knot should be hidden within the foliage and the cord should magically appear from the canopy of the plant creating the illusion of floating.

8 Suspend and display your plants as you wish. A mixture of suspended kokedama and some placed in stands creates a stunning mini citrus orchard.

The ancient art of plants

We could learn much more about the decorative potential of plants from the reverence given to indoor cultivation in the ancient East. The Japanese art of bonsai and the Chinese practice of Penjing (from which it evolved) show how to maintain a specimen tree in miniature using precise interventions such as root pruning and foliage reduction. The Vietnamese practice of Hòn Nôn Bô, creating perfect tiny replicas of natural landscapes such as islands, forests or mountainsides, was used to decorate the entrances of traditional homes. The Japanese scaled this down further with the art of Saikei or 'tray gardens' recreating entire miniature landscapes on ornamental ceramic trays. These techniques are as much about the journey as the finished article. Devotees can take years creating their labours of love and finding peace, satisfaction and fulfilment on the way. Though some of these art forms may seem extreme, consider how ornamental our plants could be if given the chance. It is my opinion that we can learn from these arts how we coexist with plants and the potential they have to be become works of art to bring beauty and joy to our homes.

MAINTENANCE

The signs that you should water your kokedama are when the moss looks dry and when the balls feel underweight. To water them, place them in a bowl of water for around 30 minutes so they have the opportunity to soak up what they need. A daily spray with a mister can extend the gap between watering. Though nutrients have been incorporated into the clay ball, citrus plants are hungry and will benefit from a specialist citrus feed added to the watering bowl from March to October when plants are flowering and developing fruit. It is worth noting that most citrus take a year to develop from blossom to fully ripe fruit. Worry not, the wait will be made infinitely more tolerable as you are treated to the most stunning visual display, as well as wave after wave of heady citrus perfume.

Edible Living Wall

Space is precious stuff and having a lot of it is often a luxury. That's true for floor space but what about the space above? Then it's all a matter of perspective. Perhaps this is partly responsible for the surge in popularity of living walls. They are the ultimate in vertical growing, turning wall space into growing space, greening every surface from office interiors to flanks of buildings and in my opinion every little bit helps. So, if you don't have room for a veg patch, why not make a veg stack? It provides a lot more planting space without taking up floor space, and it looks lush! I have 90 edible plants in mine – a huge quantity of food in a small space. This can also be a clever way to create privacy and provide a green screen to mask unsightly areas such as unattractive walls or fences.

Difficulty Level
Intermediate

You Will Need
Wall planters
Peat-free multipurpose compost
Screws
Watering system (optional but advisable)

Suitable Plants
Celery
Tomatoes
Lettuce
Strawberries
Herbs
Viola
Marigold
Pansy
Chives
Welsh onions
Chard
Beetroot (beet)

The Tools
Drill
Wooden batons (optional)

HOW TO DO IT

1 Install the wall planters as per manufacturers' instructions (see suppliers page 150). If you are doing this directly onto a wall, you may find it easier to attach rows of wooden batons to the wall first to give something to screw the planters into.

2 Using a hose or watering can, fill the reservoirs at the bottom of the planters; if you water into the top row, the others should fill via a waterfall effect.

3 Before you put any plants into the wall, stand back to have a look at the space and decide how you are going to design it. You can place all your plants randomly to create a naturalised effect, or you can use the natural structure of the planters to make a pattern with your plants. You could plant in vertical or horizontal rows, or create a playful shape or picture. I have planted my living wall in diagonal lines.

4 Plant up your living wall using peat-free multipurpose compost. For maximum visual impact mix up the plants of different textures and colours so each stands out against its neighbours. I started with upright baby celery plants followed by patio tomatoes. Next is a row of frilly lettuce offset by a bright stripe of violas, followed by strawberries, sunny marigolds, mixed herbs, a rainbow of pansies, spiky chives, bunching onions, beetroot (beet) and finally bright pink, peppermint chard. That's a huge quantity and variety of crops growing in a tiny space.

5 Once all the planters are filled, give the wall another really good soak to establish the plants in their new homes.

6 I recommend installing the optional irrigation system into the green wall. As the plants are stacked vertically, they won't capture as much rainfall as if they were planted horizontally and will need watering by hand every day in summer. An irrigation system will automatically do this by dripping water in a constant or timed basis. Install the watering system as per the manufacturers' instructions, running it along every third row to provide sufficient coverage.

MAINTENANCE

Watering is a daily necessity in the warmer months without an irrigation system. Even with an irrigation system, spot-check the compost regularly for dryness, especially when it's hot. Add a liquid feed every two weeks to help your vertical veg garden burst with life and goodies.

Saving unripe tomatoes and peppers

The sad fact of growing outside in a temperate region is that often the season is just not quite long enough to ripen all of those heat loving crops like (bell) peppers, chillies and tomatoes. It is not uncommon to find the season turning and the weather quickly cooling whilst plants are still loaded with gorgeous but still very green fruit. There is a trick to cheat the heat and force underripe crops to fulfil their full potential and it comes in the form of the humble banana. Many fruits release ethylene gas as they ripen, but the banana really pumps the stuff out more than most. It's totally natural and harmless but what it does do is accelerate the ripening process in any nearby fruit by breaking down the cell walls and converting starches into all-important sugars. We can harness this phenomenon to our advantage. Place all your unripe fruit in a shoe box or similar, if you don't have a box, a bowl or dish will do but you need to be able to cover it. Place a few bananas on top of the unripe (bell) peppers or tomatoes, the riper the better so don't chuck those wrinkly black ones that are way past their best, this is their time to shine. Put a lid on the box or cover the dish, this helps keep all that clever ethylene inside and makes sure the fruit gets a good long bath in it. Place the bowl or dish somewhere dry at room temperature and wait. Check every few days to make sure there are no signs of spoiling and remove anything that looks like it is heading that way. Before long you should start to see a change in colour of the unripe fruit as the bananas work their magic. The whole process should take around 2–3 weeks depending on just how far the fruit was from ripeness. As long as there are no signs of mould or spoiling you can just keep on going until you get as much of the early harvest to ripen as possible.

Hydroponics

‘I can’t grow plants, I kill everything!’ I hear it all the time, people believing they are incapable of keeping a plant alive. If that sounds like you, I promise you are not cursed, you don’t emit a magical plant-killing death-ray. It’s more likely that your lifestyle does not fit well with the time-sensitive demands of plant parenthood. You may have little time to water or feed consistently, and when you do turn the focus of your love on your plants, undue guilt or uncertainty may result in overwatering and excessive feeding.

Well maybe there’s a way to (nearly) remove that margin of error from the equation: i.e. you. Hydroponics is a method of growing plants without any soil. The roots of the plant are suspended in a water source, usually a moving one, that contains a balance of minerals and nutrients that provide precisely what the plant needs. It is hard to know the real nutrient levels available in compost and some can become locked up due to environmental factors. With hydroponic growing, nutrient and pH levels are both known and precisely measured, and can be tailored to individual crops to produce optimum benefits.

Hydroponic methods have become very popular recently, driven by two main factors. Firstly, these methods bypass the problems of soil-based pests and diseases, the limitations on growing space and restrictions of seasonality. It also helps that a myriad high-design domestic units have come to market aimed at time-poor and space-poor homegrowers.

Difficulty Level
Easy to intermediate

Hydroponics is finally getting the respect it deserves as a future-proof growing method – commercially and at home – and no longer a euphemism for clandestine cannabis production.

There is a plethora of units available to home-hydroponics newbies, from smart pots for one plant, to table gardens, to floor-to-ceiling smart gardens providing nearly all your fresh crops. All are intended to make growing food as easy as possible and come with everything you need to start: some you literally add water, plug in, and within days green shoots appear. Prices may seem high for the larger units, but these do provide an entire kitchen garden in your home. The price of land to grow the equivalent crops would be substantially greater.

So why is this a game-changer for those who can't grow anything? Well it's almost entirely automated. Watering and feeding are out of your hands and most units provide the optimum quality and length of full spectrum light on timers (meaning you are no longer a slave to the seasons and can grow ample food year-round). This generally means a much shorter time from sowing to harvesting as plants mature faster in perfect conditions.

What can be grown in this way: pretty much anything. All leafy greens flourish in these units and can easily provide all the salad and greens you can eat. Tomatoes will thrive in here but select patio or tumbling varieties to keep at a manageable size. Chillies, (bell) peppers, aubergines (eggplant) and even cucumbers will all grow well in these conditions. Why not try herbs and edible flowers for a beautiful and aromatic display? Even dwarf peas and beans can work well. The only thing that can be tricky to grow is root crops.

Once you've mastered the staples you can become adventurous. I have grown watercress, kohl rabi and even pepino melons. Why not stick in a dwarf sunflower just for fun? Don't be afraid to experiment, this is a versatile method and if one crop doesn't work you can pull it out and try a new one. Enjoy it!

MAINTENANCE

Keep an eye on water levels and top up where necessary. Some units may require you to measure the nutrient and pH levels and adjust accordingly: get in the habit of doing this on a weekly basis to make sure everything stays on track.

Future-proof farming

Thanks to climate change caused by careless human activity our planet is becoming a more challenging environment. Even as we try to curb our impact on this wonderful world, there is no escaping the consequences of what we have done and the problematic growing conditions that follow. Innovation in commercial farming is necessary to reduce the carbon footprint of our food, but also to evolve as the environment changes to be more in tune with it. Hydroponics is one method that has been used to future-proof our fresh food supply. Offering a way to grow completely vertically, it vastly reduces the footprint of a growing space with huge quantities of crops on vertical towers with a base of under one metre square. These can be placed in temperature-controlled indoor environments where the outside world is no longer a factor and artificial lighting means that growing is seasonless. Many of these farms can be run off-grid, capitalising on renewable energy, and located to make minimal ecological impact. These freedoms have resulted in a food-growing renaissance, from microgreen farms in abandoned London tunnels, to lush greenhouses in the Arizona desert, to farms providing fresh green produce in the Arctic Circle permafrost. No longer does a farm require vast acreage, heavy machinery and annual crop rotations. Almost any space can be exploited for growing. What starts in industry inevitably filters into our homes, and embracing innovation, more efficient techniques and technology is the way towards a healthier, more self-sufficient life.

CLICK GROW

YE
6
No 10575

Farm Desk

Difficulty Level
Intermediate

You Will Need
Customisable metal shelving
Hanging cutlery pots
Shallow trays
Full spectrum grow lights
Timer switch
Hemp microgreen matting

Suitable Plants
Microgreens
Vegetable plants
Edible houseplants

The Tools
Mallet
Cable ties

Thanks to flexi hours, self-employment and the increasingly open minds of many employers, more of us are working from home. The key to a happy and productive home-working environment is a soothing and harmonious workstation. There is a theory in architecture called biophilic design that essentially aims to integrate the natural environment into interior spaces. It is based on the idea that humans have an ingrained need to connect with the natural world due to the fact that we have evolved to depend on it for our very survival. This is why we naturally seek to spend our recreational time in green spaces such as gardens, parks and the countryside – the natural world relaxes us, leaving us feeling calm and fulfilled. The integration of the natural world into our home environment and, more specifically, our working environment is thought to benefit productivity, stress levels, health and wellbeing. What better way to integrate nature into your work life and your home than to create a dreamy botanical workstation that doubles as a home farm?

N.A.A.F.I

HOW TO DO IT

1 Chrome wire shelving is easy to get hold of (see suppliers page 150) and can be bought in wide and narrow sizes and half and full height. This makes it almost infinitely customisable and can be tailored to the space it is destined to occupy. For a small space, aim for the narrower shelves; for larger areas of wall, you can use the longer shelving units. Just make sure that you get at least one of the longer shelving units, this will form the actual desk part of your indoor farm.

2 Make sure to build your desk from the bottom up, adding in the base shelves first and working your way up. Have a mallet handy to make alterations as you go. Do not add any base shelves in the space that will form your desk, so you retain somewhere to work! The average height for a desk is 73.5 cm (29 in) but for the ultimate ergonomic workspace, enter your height into an online desk-height calculator for your perfectly comfortable working position.

3 Shelves can be spaced at intervals to suit your requirements whether it is files, books or stationery. Just bear in mind that you will need a minimum shelf height of 30 cm (12 in) for sufficient room for plants. I like to include a few higher shelves to accommodate larger plants, variety is after all the spice of life.

4 Create at least one shelf with a height of 15 cm (6 in), this will be your microgreen area. Use cable ties to attach full spectrum grow lights to the underside of the top shelf. This will allow you to grow microgreens intensively in limited space regardless of natural light levels. (See Microgreens project, page 61, for how to grow microgreens in trays on hemp matting.)

5 Now you are ready to start filling up your shelves and the choices are limitless. Fill the farm desk with herbs, fruit, vegetables and edible houseplants. Nestle troughs of baby kale alongside your files, pots of herbs and chillies can form fabulous book ends. Let tomatoes tumble from the shelves while oxalis, fuchsia and aubergines (eggplant) flourish amidst the stationery. You can add even more growing space using hanging cutlery pots spilling with chard or dwarf beans. The lattice work on the shelves becomes a kind of trellis to grow climbers such as passion fruit and beans around the frame.

6 If you are not able to position your farm desk adjacent to a window, don't be afraid to add a few clip-on grow lights to give your farm a light boost and ensure you can continue to grow through the winter.

MAINTENANCE

The plants will need feeding and watering regularly. Place all lights on a timer set to be on from 7am to 9pm to remove the need to turn them on and off. Make sure the hemp mats do not dry out to maintain a good crop of greens; once all the greens in one tray have been harvested you can re-sow on fresh matting to ensure a continuous supply. Though they are single use, hemp mats are compostable and biodegradable.

Off with its head!

Want to know the secret of how to grow big, bushy and super-productive chilli and (bell) pepper plants? It sounds crazy but you need to chop off their heads! Plants contain a growth hormone called auxin, and on a seedling the auxin is concentrated in the growing tip, the top few centimeters (inch). Left to its own devices, this seedling will keep growing upwards in a straight line, but we want it to branch out all the way down the stem because more branches mean more flowers which ultimately means more fruit. It may seem brutal, but if you chop off the top 5 cm (2 in) of the seedling you will cause the plant to redistribute its auxin to multiple places further down the stem. For a week or so you may not notice anything, but before long little green lumps will appear at most if not all of the nodes (just above where the leaves are attached to the stem). Each of the nodes will quickly become a new growing tip. Meet your new, super-bushy, multi-branched pepper plant capable of significantly higher yields. A little tough love goes a long way!

Edible Terrarium

Difficulty Level
Tricky but worth it

You Will Need
Glass display cabinet
Clear silicone bathroom sealant
Full spectrum sunlight grow light (not pink or red tint)
Double-sided sticky pads
Expanded clay pebbles
Sphagnum moss
Activated charcoal
Peat-free multipurpose compost
Driftwood or mangrove root
Ornamental rocks
Twine
Cushion moss
Chamomile tea bags

Suitable Plants
Vanilla orchid
Pineapple
Purple shamrock

Non-edible Extras
Spiderwort (Tradescantia)
Nerve plant (Fittonia)
Tufted airplant (Guzmania)
Elephant ears (Alocasia)

We are all now conscious of the impact our lives and how we live them have on this precious planet, and most of us strive to reduce that impact. Growing some of your own food is a giant step in that direction, but what if you have a penchant for the exotic? Delicious tropical treats such as vanilla and pineapple can't easily be grown in temperate climates and are flown in from afar. But now you can create your own slice of jungle, turning a glass display cabinet into a gorgeous giant terrarium. The hot and humid ecosystem inside is the perfect environment for growing tasty tropical plants, while also providing a jaw-dropping feature.

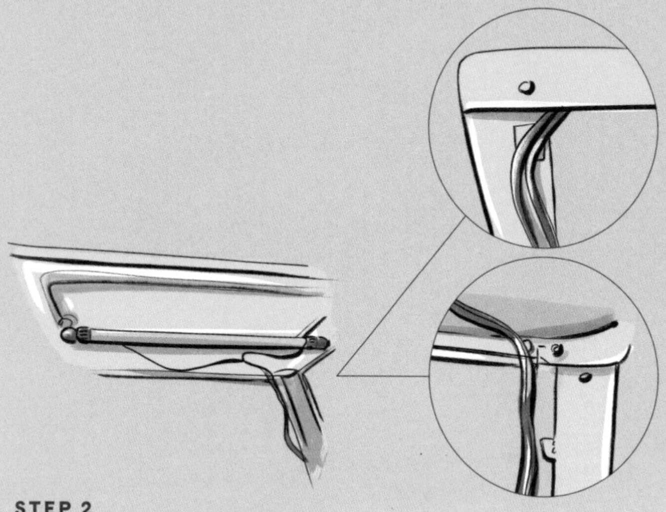

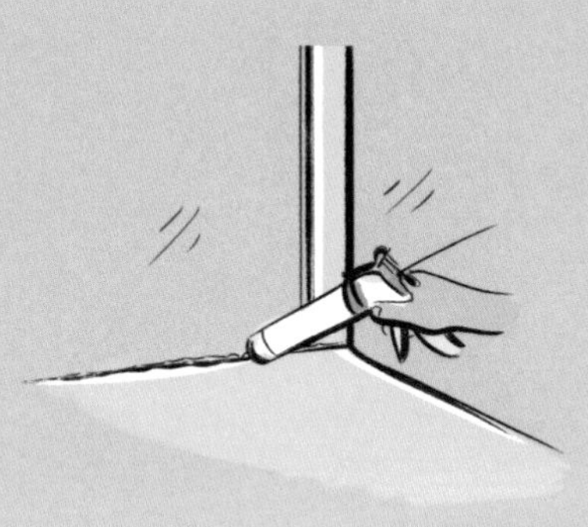

STEP 2
Attach your light to the roof of the cabinet. The wires should run out of the back corner, sandwiched between the back pane of glass and the lid. Use sticky fixers to secure the wires out of sight.

STEP 3
Make the bottom section of your terrarium watertight.

STEP 4
Use a glass shelf to make a watertight planting section.

STEPS 5–8
Create the base to plant into.

From top to bottom:

Planting layer:
12 cm (5 in) peat-free multipurpose compost.

Purifying layer:
200–300 g (7–10½ oz) activated charcoal.

Filter layer:
Thin layer of sphagnum moss.

Drainage layer:
Expanded clay pebbles, 3 pebbles deep.

HOW TO DO IT

1 Build your cabinet as per manufacturers' instructions, but don't attach the top or the internal shelves which you should put to one side for now. Fixing the grow light into the top before installing it will make life a whole lot easier!

2 Install your grow light in the top of your terrarium. This will need a 14-watt bulb to sustain healthy plant growth. Assemble it following the manufacturers' instructions and plug it in to check it works. Attach your light to the roof of the cabinet using double-sided sticky pads (many lights now come with these). Now attach the top onto the cabinet as per manufacturers' instructions, you want the wires of the grow light to run out of the cabinet in one of the back corners, sandwiched into the gap between the back pane of glass and the lid. Use sticky fixers to secure the wires out of sight to the back corner of the cabinet.

3 To make the bottom section of your cabinet watertight, use the silicone sealant to seal all around the base and up the sides to a height of 20 cm (8 in).

4 Next take one of the internal glass shelves (that you put aside) to make the front part of the planting section of your terrarium. Place the shelf on one of the long sides and place it against the inside of the door frame so that the bottom third of your terrarium now forms a glass tank. Use the silicone sealant to stick that in place, filling gaps and making it watertight, and leave it to dry for several hours.

5 Now that you have created the framework for your jungle, it is time to create the base. Cover the bottom of the terrarium with expanded clay pebbles to around three pebbles deep. This layer is for drainage and where excess water collects to avoid drowning the roots of your plants.

6 Cover the pebbles with a thin layer of sphagnum moss. Spread it thinly, if you have a few gaps that's fine. This will act as a filter to stop soil falling and filling up the spaces in your drainage layer.

7 Sprinkle activated charcoal powder across the surface of the sphagnum moss, about 200–300 g (7–10½ oz) will do it. Charcoal has a purifying effect and will help to keep your terrarium clean.

8 Add a good layer of peat-free multipurpose compost for the planting layer: firm it down a bit to get rid of air pockets but don't stomp it down hard. If it's too compact your plant roots won't be able to breath. Once firmed down this layer should be 12 cm (5 in) deep.

Propagating pineapples

Once established, this fruiting bromeliad will send up baby plants, called pups, from its roots around the base. Use a clean sharp knife to slice off these pups from the plant's main roots and place them into potting compost, keeping them warm and moist. They will quickly grow roots of their own to become new pineapple babies! You can actually grow a brand-new plant from the fruit itself. Cut the spiky top off your pineapple and any remaining fruit, leaving just the leafy part. Strip away the bottom leaves until you have around 1 cm (½ in) of exposed white stem at the base. Balance the leafy top on the rim of a jar filled with water so that the stem is submerged. Place in a bright spot but out of direct sunlight and make sure the water stays fresh and topped up. In around two weeks little roots will start to grow out of the stem. Plant in potting compost to grow on or better yet, into your own terrarium.

9 Add a large piece of driftwood or mangrove root. These are available from terrarium suppliers or reptile pet shops. When selecting your wood, make sure you measure the cabinet's internal dimensions to ensure it will fill the space and still fit inside. Look for an interesting twisted piece as a focal point and for the vanilla orchid to climb. Complete the look with a few pieces of rock placed on the compost to break up the green carpet of plants on the terrarium floor: I chose one large and two smaller pieces.

10 Now you can start filling your terrarium with tropical plants. A good terrarium looks like a slice from a thriving natural environment, and to achieve this you may need to mix non-edible plants with your tropical edibles – but keep the edibles as the focal point of this living piece of art. Vanilla Orchid is a climbing plant so should be planted at the base of your wooden centrepiece. Create a 'forest floor' feel with baby pineapple plants and purple shamrock, mixed with non-edible spiderwort (Tradescantia), nerve plant (Fittonia), tufted air plant (Guzmania) planted in and around the rocks and base of the wood. Height at the back of the display will enhance the jungly vibe so here I used two smaller varieties of non-edible elephant ears (Alocasia).

11 To complete the jungle effect, add a few plants in the nooks and crannies by creating wabi kusa balls. These are basically little kokedama where the roots of the plants are encased in clay and wrapped in sphagnum moss (see Kokedama project, page 93). These wabi kusa balls should be lodged into the wood's recesses, producing thriving aerial plants, each with its own life-giving earth and reservoir.

12 Give all your plants a good soak but stop before water pools in the drainage pebbles.

13 Cover any exposed compost with moss mounds to add the finishing touches to your slice of jungle.

14 Finally, brew up a strong batch of chamomile tea and allow it to cool. Place it in a mister and spray the whole of this inside your terrarium including on the foliage of your plants. Chamomile contains good levels of sulphur and has a natural anti-fungal action to help keep your terrarium free from mould and mildew.

MAINTENANCE

A terrarium is a mini, self-sustaining ecosystem which needs relatively little maintenance. Position it away from windows and direct sunlight so the grow lights can manipulate the light inside as per the plants' needs. You will find you don't need to water it often, particularly after a couple of months when established.
A successful terrarium will have its own water cycle going on inside. Rather than adopting a regular watering routine, watch the planting layer for signs it needs water: the compost will change colour as it dries, becoming lighter, and indicating a need for water. If you see water pooled around the pebbles in the drainage layer, the environment has more water than it needs, so don't add more for a while. Another thing to consider is the addition of springtails to your little ecosystem. Though not essential, these little insects eat dead leafmatter and will act as an in-house cleaning crew.

STEP 9
Add a large piece of driftwood or mangrove root.

STEPS 10–11
Fill your terrarium with tropical plant babies. Add plants in the nooks and crannies using Wabi Kusa balls, see page 93 for method.

Indoor Trellis

Difficulty Level
Intermediate

You Will Need
20 cm (8 in) bamboo dreamcatcher craft rings
2mm (½ in) anodised aluminium wire (many colours to suit)

Suitable Plants
Kiwi 'Jenny' *(Actinidia deliciosa)*
Kiwi berry 'Issai' *(Actinidia arguta)*
Passion fruits *(Passiflora caerulea* or *Passiflora edulis)*
Mashua *(Tropaeolum tuberosum)*
Runner bean 'Sunset'

The Tools
Wire cutters
Drill
Pliers
Partially threaded screws
Rawplugs (optional)

There are so many incredible edible climbing plants! Perennial fruits like kiwi and passion fruit make fabulous houseplants, and once established should provide a crop of fruit year after year. Mashua, a perennial nasturtium, offers an abundance of peppery leaves for salads and pesto, along with its stunning, trumpet-shaped, edible flowers, as well as edible underground tubers. Annual vegetables such as beans are just as visually striking: try 'Sunset' runner beans for a wall of delicate peachy flowers and a glut of delicious vegetables.

However, growing climbing plants indoors comes with a challenge, they need something to climb. This is an easy-to-make trellis to fit any available wall space, like a living piece of art. Kiwis in the bedroom? Why not. If you have a wall space, no matter how uniquely shaped, you can create a bespoke and beautiful trellis to fit and fill your wall with green. All the materials needed are widely available from crafting suppliers.

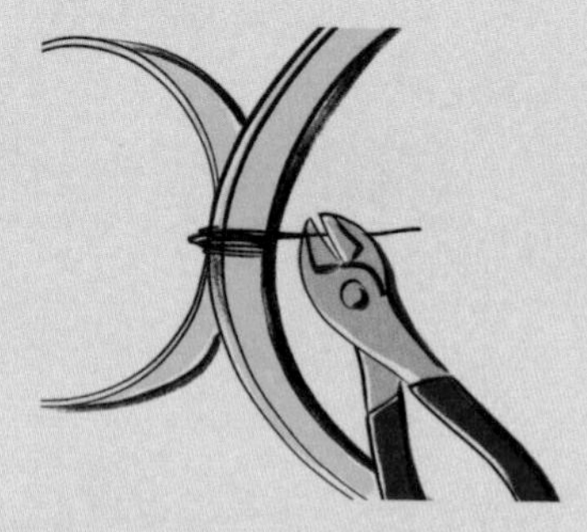

STEP 1
Wrap aluminum wire around two of the bamboo rings 4 times, cut and tuck ends in.

STEPS 2–3
Get creative. Continue to attach more rings to make whatever shaped trellis you want.

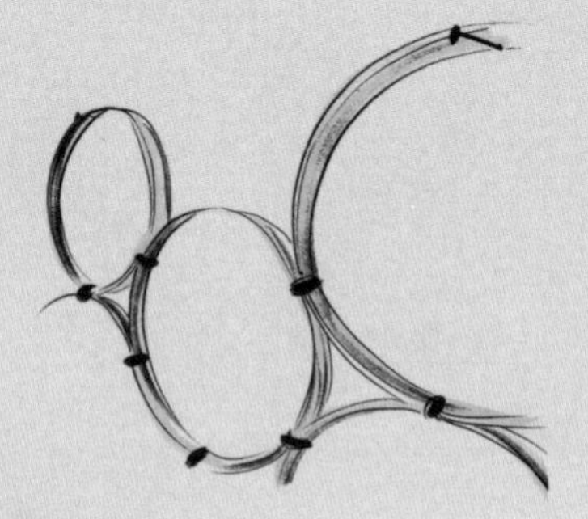

STEP 4
Hang your trellis using screws in the wall.

STEP 6
Place your plant in the best position on the wall. Wind longer stems in and around the trellis to get it started in the right direction.

HOW TO DO IT

1 Take two of the bamboo rings and place the flat outer edges together. Use the aluminium wire to tightly bind the place where the two rings touch, winding it round four times. Aluminium is light and flexible so despite its thickness it's easy to manipulate. Using the wire cutters, cut the ends off close to the rings and use a pair of pliers to gently tuck the ends in against the rings. The side that has the ends visible is the back and will go against the wall.

2 Now take another ring and place it so that the three rings form a triangle. Use the same binding technique to attach this third ring in the two places it touches, making sure the wire ends are at the back.

3 Now it's time to get creative: you can attach more and more rings to make whatever shape trellis you want. You can create an organic shape that flows up and across the wall, or, like me you can use a wall-mounted planter and envelop it with trellis. For a more formal look you can create a geometric design like a square or rectangle to fill the space.

4 Another pair of hands will make it easier to hang your finished trellis but it's not vital. Hold your trellis against the wall in its final position, and mark the two highest points on the wall, with tiny pencil crosses, inside the rings: this is where you will place the two screws to keep your trellis steady. Put your trellis to one side and using the drill, insert screws where you have made your marks, using rawlplugs if necessary. It's neater to use partially threaded screws as the screws should protrude around 2 cm (¾ in) from the wall.

5 Hang your trellis on the screws and check if you need any additional fixings to keep the trellis stable. The trellis should hang, not be screwed up tight against the wall, to give your climbers space to weave through it as they grow, however you do not want it swinging around wildly so some additional screws may be prudent.

6 Place the container holding your chosen plant in the best position for it to climb. If your plant already has some longer stems you can begin training these in and around the trellis to get it started in the right direction.

MAINTENANCE

Regular watering and feed will keep your climber heading up and out across the wall. Train in any unruly stems that grow in the wrong direction, remember you are the boss. Within reason you can make the plant grow into whatever shape you desire.

Fresh take

Trellis-making works for other shapes and materials too. You can use your new wire-binding skills with macramé rings of different shapes, each shape producing a different pattern; or you can mix shapes within one design to produce a complex geometric design. Metal rings (rather than bamboo) will produce a modern, industrial look. Why not try using rings of varying sizes for a highly abstract effect?

Cutlery Pot Vertical Veg Garden

Difficulty Level
Intermediate

You Will Need
Six 57 cm (22 in) kitchen storage rails
18 hanging cutlery pots
36 hooks to fit rails
Three 180 x 20 cm (70 x 8 in) wooden planks, a minimum of 3 cm (1 in) thick
Three wooden batons 55 x 3 x 4 cm (22 x 1½ x 2 in)
Peat-free multipurpose compost

Suitable Plants
Celery
Patio tomatoes
Nasturtiums
Lettuce
Chard
Chilli

The Tools
Drill
3 mm (½ in) drill bit
Screwdriver drill bit
60 mm (2½ in) screws
Pencil
Tape measure

Do you dream of having your own kitchen garden? Of wandering among rows of lovingly tended crops, with delicious, diverse harvests of fresh, organic vegetables? That's a beautiful dream of the good life, but many people either have little outside space or none at all. Well, here is a gorgeous way to grow those rows of crops in your home, year-round, and it uses minimal floor space. This mobile, vertical garden is mounted onto a backboard that you can move as you wish, from balcony to garden, anywhere with light. You can use any strong wood planks for the backboard (providing they are 3 cm (1 in) thick) but reclaimed boards are most striking. I used planks from a local carpenter – their fabulous knots and cracks made them unsuitable for furniture but perfect for my background to this vertical garden. Scaffolding boards are a great alternative and widely available from reclamation yards or online. For a cleaner, more modern look, larger hardware stores will have pine boards in the correct sizes and will often cut them to length. This project is customisable so you can paint or stain them for the look you want, let your creativity run wild!

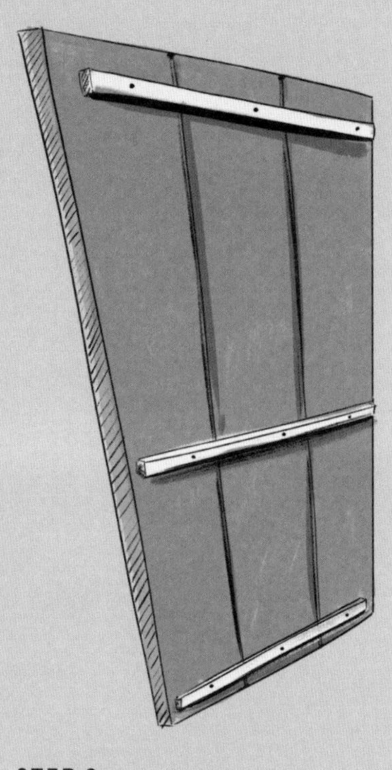

STEP 2
Place your batons on the back board planks at right angles to the direction the planks are lying. One should be located across the middle of the planks and the other two should be located 10 cm (4 in) from each end of the planks.

STEP 6
Install your first rail as per manufacturers' instructions, so that it is right at the very top of the back board.

STEP 9
Fill with compost and plant celery, a row of lettuces, chilies, a row of tomatoes, peppermint chard that adds a striking flash of neon pink, and finally a beautiful cascade of nasturtiums on the bottom row.

HOW TO DO IT

1 Take a moment to examine the boards and choose the best side for the fronts: for a rustic look this may be the side with knots and markings, for a clean, modern look or if you plan to paint them, this may be the flawless side. Once you've chosen the fronts, place each board face down, side by side. Align the ends with the sides touching all the way down.

2 Place the batons on the backboard planks at right angles to the direction the planks are lying. One should be placed across the middle of the planks and the other two 10 cm (4 in) from each end.

3 Using the drill and the 3 mm (½ in) drill bit, drill a guide hole through the baton and into the middle of each plank, this will make it much easier to get the screws in and will ensure that your batons don't split when you do it.

4 Switch the bit on your drill so that you now have the screwdriver head attached. Remember to use the appropriate bit, Phillips for a crosshead screw, flat for a flat-head screw. Now take your 50 mm (2 in) long screws and use the drill to screw them into the guide holes.

5 Turn over the backboard so it is now face up.

6 Install the first rail at the top of the backboard following the manufacturers' instructions. Check now whether the rails have an up and a down: the ones I used (see suppliers page 150) had little screws to secure the rails into the bracket and these needed to face down to be hidden. Additionally, note whether your hooks need to be threaded onto the rail before they are installed or can be hung later. Time spent checking now avoids the frustration of repeating steps later on.

7 Using a tape measure, measure 30 cm (12 in) down from the middle of each end of the top rail and mark with a pencil. This mark will be the middlepoint of the second rail, install it in the same way as the first rail so that the middle of the rail lines up with your pencil mark.

8 Repeat step 7, installing your rails at 30 cm (12 in) intervals all the way down the length of the backboard.

9 Now for the fun bit! Time to plant up the vegetable garden of your dreams. The joy of cutlery pots is that they usually have an integrated drainer at the bottom, which means you can fill them with compost and are good to go. For the top row, try three humble celery plants, a staple in so many dishes, and treat them as a cut-and-come-again crop, harvesting outer sticks and leaving the hearts to carry on. It gets tall so works well on the top row. Add a row of lettuces for all your salad needs followed by a row of chillies. Keep the chilli plants well pruned in this situation to help them cope with the smaller than ideal pot. A row of tomatoes is a staple next and a patio variety will stay neat and petite. For your leafy greens, grow a row of chard, such as peppermint chard, to add a flash of neon pink to your patch. Finish with a cascade of nasturtiums on the bottom row, the leaves are an easy alternative to rocket (arugula) with an incredible display of delicious flowers.

10 Add a hook to one of the rails to hang your harvesting scissors and watering can.

MAINTENANCE

Avoid overwatering these pots by waiting until the surface of the soil is dry before watering. Some of these crops such as tomatoes and (bell) peppers should ideally be grown in larger containers, so regular feeding is vital while they are flowering and fruiting. A weekly feed with organic liquid feed should ensure a long season of harvests.

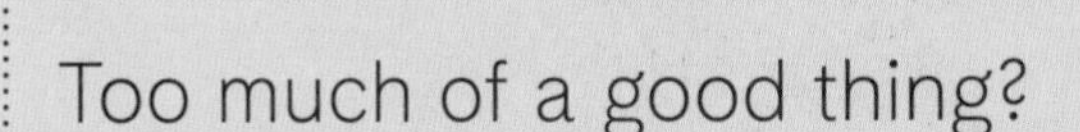

Too much of a good thing?

So, you had a really successful season, your plants are thriving, and you are drowning in more home-grown goodies than you know how to handle. What do you do? Preserve it of course! There's nothing more satisfying than eating home-grown produce in winter. But no need to crack out the frilly apron and embrace your inner Stepford wife: personally, I can't make chutney to save my life, I won't pickle, and jam-making (jelly-making) is a dark art I'll never master. I do however preserve excess harvest in vast quantities, quickly and with minimal effort. How? The freezer is your best friend! Many crops freeze perfectly after a two-minute blanch in boiling water, for example leafy greens, root crops, aubergine (eggplant), beans and peas. Once blanched, allow them to cool, dry them a little and then pop them into deep freeze until needed. Even better, tomatoes, (bell) peppers and herbs can go straight into the freezer. Just bear in mind that the action of freezing causes some cell structures to collapse, meaning they become mushy when defrosted. They won't be good for eating fresh but are perfect for adding to cooked dishes, retaining all their original flavour and nutrients.

Indoor Greenhouse

Light, moisture and warmth are a plant's basic needs but if you boost these factors you can really stimulate productivity. Gardeners have long known that the best way to do this is with a greenhouse. The glass amplifies light while the frame traps warmth inside and creates a stable, humid microclimate that plants love. Many fruit-bearing staples originate from warmer climates and benefit from this environment, such as tomatoes, (bell) peppers, chillies, cucumbers and aubergines (eggplant). But many gardeners don't have space for the greenhouse of their dreams and some have no outside space at all. Well, it is easy to simulate a small, warm growing environment indoors using a glass display cabinet, plus it's undeniably gorgeous. You can repurpose any display cabinet into a greenhouse if it has glass at the back and front. That way, when placed in front of a window, the sun's light and warmth will create a perfect microclimate for propagating plants, maximising yields of fruiting crops and indulging you with adventurous tropical varieties.

Difficulty Level
Easy

You Will Need
Glass display cabinet
 with a glass back
Full spectrum grow lights

Suitable Plants
Tomatoes
Patio cucumbers
Aubergines (eggplant)
(Bell) peppers
Chillies
Tropical plant species
Plants in propagation stage

Sow seeds like a rebel

Call me a rebel but I'm financially savvy with my propagating. The traditional method for sowing seed says we should initially use seed and cutting compost, then when large enough we should pot up the baby plants into potting compost, finally rehoming them into standard multipurpose compost at stage three. The reason for this is that for the first few weeks, a seed has its own food supply containing everything it needs, until it grows roots and draws nutrition from the soil. Seed and cutting compost is close to sterile with little to no nutrient content. Potting compost has higher levels of nutrients which is why you move the plant baby onto this as it begins to grow. Multipurpose should be a fully nutritionally balanced growing medium so, as the plant baby matures, placing it in this medium will provide everything it needs. So, here's my thinking... just because a newly germinated plant will not be drawing any nutrition from the soil, shouldn't mean it can't be there anyway ready for when it does need it. I skip two steps and sow all my seed directly into multipurpose compost. This means I save money on the specialist propagation composts and can often go longer between potting up, taking my cue from the roots as they outgrow the container, as opposed to the plants' needs for additional nutrients. Pretty rebellious stuff I know, so, are you ready to be a propagation provocateur?

HOW TO DO IT

MAINTENANCE

Open the cabinet doors for a few minutes every day to allow fresh air to circulate. Plants inside can dry out quickly – especially at the side nearest the window – so check closely whether more watering is required. Grow lights can be timed to turn on and off automatically, I set mine to summer daylight hours, on at 7am and off at 9pm.

1 Assemble your cabinet as per manufacturers' instructions without installing the shelves.

2 Position the cabinet in front of a window where it can harness maximum sunlight. The larger the window the better, and in the Northern Hemisphere this will be a south- or west-facing window which experiences the most intense sunlight for the longest period. An east-facing window will receive only morning sun which is generally not as intense, and a north-facing location simply won't receive enough light to benefit. Reverse this for the Southern Hemisphere.

3 The positioning of the interior shelves will be dictated by what you want to grow: if you want to propagate plants the shelves can be close together, but mature chilli and (bell) pepper plants need more head room. Personally, I like to mix it up, with shallow shelves for propagation at the top, mid-height shelves for low-growing plants below them, and a large space for mature specimens at the bottom.

4 Light is the key for a productive greenhouse. However, light intensity decreases in the winter, and the hours of daylight reduce. This means that your plants may enter a period of dormancy, with annual plants dying back. Additionally, the second half of winter is a good time to start propagation and an indoor greenhouse allows a quick start ahead of spring. Young seedlings especially require good light levels to stay healthy or they grow tall, thin and leggy, with weaker and shorter lives. Poor winter light levels can be conquered by installing grow lights inside the cabinet and enabling year-round cultivation. Fix them centrally above each shelf with strong double-sided sticky fixers, so the light is evenly distributed.

Edible Art

When I say it is possible to grow food artistically in your home, I'm not joking. Everyday herbs can be elevated to the status of living art through a clever display and an understanding of habit. Many herbs have a creeping habit meaning they grow to make a flat, living mat rather than growing tall. Corsican mint is a micro-leaved, mat-forming herb with a pungent peppermint aroma. Many fabulous thymes such as 'Lemon Curd', 'Pink Ripple' and 'Carbon Wine and Roses' creep and put on annual displays of bright pink and purple edible flowers. Prostrate rosemary develops beautiful snaking, sculptural tendrils. Chamomile, famous for its scented lawns, is a non-flowering, feathery, aromatic herb that is still used to make relaxing tea. Alpine strawberries are low growing plants that will produce miniature super sweet fruit. All these plants are a treat for the nose as well as the eyes, filling a room with intoxicating scents, and their mat-forming habit makes them ideal to frame on a wall like a picture. With some planning and a little modification, you can create living art works that evolve in your home, while also offering year-round harvests.

Difficulty Level
Tricky but worth it

You Will Need
- Box picture frame (made-to-measure, order online)
- Beeswax polish
- Mangrove root or driftwood
- Large gauge chicken wire
- Sphagnum moss
- Peat-free multipurpose compost

Suitable Plants in 9 cm (3½ in) Pots
- Creeping thymes
- Prostrate rosemary
- Corsican mint
- Chamomile lawn *(Chamaemelum nobile)*
- Alpine strawberry

The Tools
- Glue
- Wire
- Wire cutters
- Saw (to cut driftwood)

STEP 2
Glue the corners of frame inner slips together to form a box.

STEP 3
Coat the inside of your frame with beeswax polish.

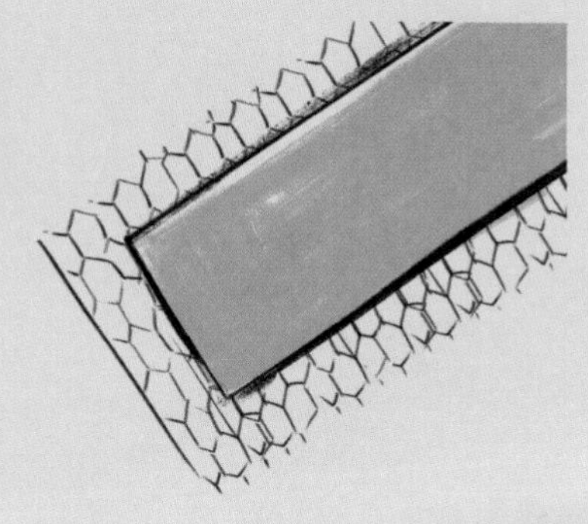

STEP 4
Cut a section of chicken wire, 5 cm (2 in) larger on all sides than the backboard of your frame.

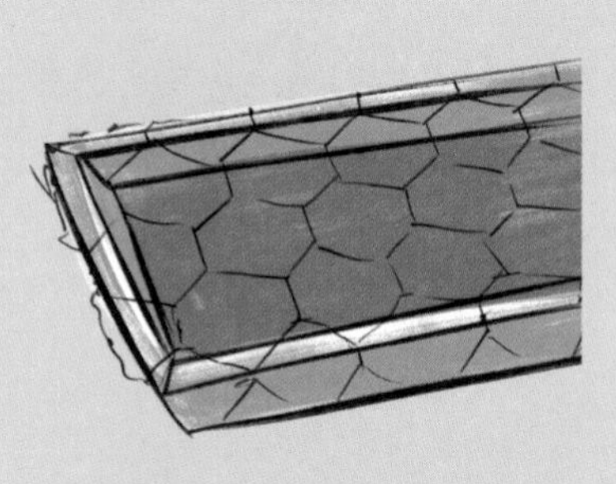

STEP 5
Place the chicken wire section on top of the slips and bend the wire over and around the slip box.

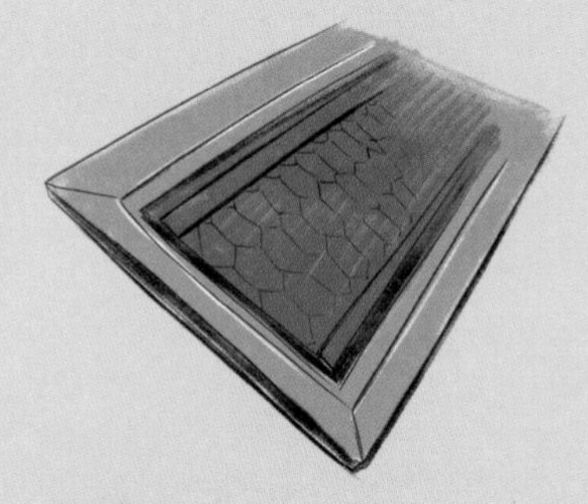

STEP 6
Turn the slip box over and insert it into the frame, wire covered side first.

STEP 7
Arrange the driftwood or mangrove root on the front of your frame and wire the wood onto the chicken wire across the front of the frame.

HOW TO DO IT

1 Order a picture frame from an online made-to-measure service. This gives enormous flexibility over scale and shape, and like my series of three frames here, you can be creative in the shapes you order, but remember that more extreme shapes will make the next steps trickier. Whatever its shape, request your frame to be 7 cm (3 in) deep, with a plexiglass front, and most importantly, that its 'inner slips are 10 mm (½ in) shorter on all sides so they fit loosely within the frame'. This is because standard frames come with snug-fitting 'slips' inside which make a smaller frame to mount the picture. We are going to wrap our slips with chicken wire, so we need the slips to be less snug than standard.

2 When your frame arrives, glue the corners of the inner slips together to form a box.

3 While this dries, coat the inside of your frame with beeswax polish, getting into all the corners and protecting the wood from future damp conditions.

4 Spread the chicken wire out and, using the backboard of your frame as a template, use the wire cutters to cut a section that is 5 cm (2 in) larger on all sides than the board.

5 Place the chicken wire section on top of the slips and carefully start to bend the wire over and around the slip box. Keep going until the wire fits fairly tightly over the slips.

6 Turn the slip box over and insert it into the frame, wire-covered side first. Be careful not to get the wire caught on any of the clips that hold the backboard in place, a standard dinner knife can help to bend the clips back so the wire can slide past. The wire and slips should now be sitting snugly on the inside of the frame with no wire protruding.

7 Take the driftwood or mangrove root and place it where you want it on the front of your frame. You may need to cut larger sections with a saw so they fit comfortably in the frame. Don't worry if your wood overlaps the frame in places, this can look effective. Aim for one large focal piece of wood, or several smaller pieces spaced around the frame. When you have decided on their final positions, wire the wood onto the chicken wire across the front of the frame. Driftwood and mangrove root are both twisting and gnarly with nooks and crannies to run the wire through and be hidden at the front.

MAINTENANCE

Mist your edible art daily to keep it moist. Once a week remove it from the wall, lie it on a flat surface and water sparingly.
Let it rest for 10 minutes or more before hanging it back on the wall.
A monthly liquid feed will help your art thrive, and you should regularly trim the herbs to keep them prettily contained in the frame.

8 Take the creeping plants and roughly decide where to place them in the frame, mixing the textures: for example small-leaved Corsican mint next to spikey prostrate rosemary, or frilly chamomile blending into tendrils of creeping thyme. One by one, remove each plant from its pot and, saving it into a large bowl, carefully shake off excess soil from the roots so that they are small enough to fit through the holes in the chicken wire. Gently push the herb roots through the chicken wire: if this is a struggle you can use the wire cutters to snip a filament of chicken wire to make a hole larger, remembering to bend the wire back once the plant is in position. Keep doing this until all the herbs are in the frame.

9 Turn the whole frame over. Take the sphagnum moss and add a 1–2 cm (½–¾ in) layer, filling between the plants, packing the moss at the base of each plant so its roots can't escape the frame.

10 Pour all the compost saved in step 8 into the frame and top up with peat-free compost, gently pressing it down until the whole of the slip section is back filled. Give it a water but don't soak it, you don't want water pouring out of the front of the frame.

11 Take the plexiglass front, remove its protective film and fit it into the back of the frame. Do not use the backboard, it is made of chipboard and will absorb water from the compost and distort. Fold down the clips on the back to secure it in place, the dinner knife can make this easier on the fingers.

12 Add a secure method of hanging to the back of the frame such as a loop or picture wire, just remember this frame is heavier than a standard picture.

13 Now you can display your living art on the wall.

Make the most of your chamomile 'lawn'

As well as sporting stunning foliage and an intoxicating aroma, chamomile *(Chamaemelum nobile)* is a useful plant to have around. The leaves of this non-flowering lawn plant can be used to make a soothing bedtime brew. Place a sprig or two of fresh chamomile foliage in a cup and pour over some hot water. Allow it to steep for a couple of minutes then remove the leaves. This tea is naturally relaxing and smoother-flavoured than traditional chamomile tea. And it's good for your plants too, but in a stronger brew. For plant tea, take lots of cuttings – you can't use too much, you want the infusion as strong as possible – and leave them to steep in hot water until cooled. When cool, transfer into a spray bottle and you have your very own organic anti-fungal plant spray.

STEP 8
Take the creeping plants and decide where to place them in the frame. Gently push the roots through the chicken wire.

STEP 9
Turn the whole frame over. Add a 1–2 cm (½–¾ in) layer of sphagnum moss and fill between the plants.

STEPS 10–11
Fill the frame with the compost, gently press it down until filled. Water it and add the plexiglass front into the back of the frame.

Eat Your Houseplants

Botanical interior styling has seen an unprecedented surge in popularity in recent years and it is spearheaded by the young and style-conscious. The benefit of filling one's home with plants is well-documented: they have a positive impact on health, productivity, creativity and general stress levels. They are even proven to benefit our physiological wellbeing, filtering harmful toxins and general pollution from the air we breathe, creating a cleaner, oxygen-rich and humidified environment. *Phytoncides*, those anti-microbial compounds released by trees and plants to protect them from rot and discourage animals from eating them, are proven to boost the human immune system for around seven days after exposure. Even the natural microbiome of the soil can positively benefit the human body. Medical studies have shown that *Mycobacterium vaccae*, a harmless soil bacteria, increases serotonin production in the brain, making people happier and better able to cope with stress, like a natural antidepressant.

The missed opportunity here is that houseplants are a source of beauty as well as tasty and inspiring food. Many popular houseplants are in fact edible – so you may already have an edible indoor jungle without knowing it. Additionally, there are many beautiful, edible alternatives to popular houseplants, so you can keep the same dreamy botanical interior and eat it too. This could be a spur to growing some exciting tropical plants that you may never have considered. You might even rethink the purchase of an imported plant from a specialist supplier, thereby avoiding the food miles of something flown in from the tropics. You can't expect instant results, some of these plants could take a long time to mature and produce a crop, but they will enrich your life and your home on the way.

Difficulty Level
Easy

Houseplants you didn't know you could eat

Elephant ears – What's in a name? Well quite a lot when you consider the humble elephant ears, a staple houseplant. There are in fact two plants that bear the common name elephant ears and so you need to pay serious attention to the Latin. *Alocasia* are probably the most commonly available elephant ears and are not to be eaten. However *Colocasia* are a staple root crop, otherwise known as taro, and just as grand, beautiful and colourful as their inedible namesake. Taro is one of the oldest cultivated crops and has been grown for thousands of years throughout Asia for its edible corms. Archaeologists have found evidence of taro as a food source from over 11,000 years ago and it remains a staple crop today in African and South Asian cultures. Even the young leaves of this amazing plant are edible after being boiled twice. As a tropical species it is not usually considered a viable crop for temperate climates, however when grown in the warmth of your home the game is changed. Instead of potting on a large and successful plant, remove some of the outer corms and put aside to add a new dimension to dinner.

Oxalis – A hugely popular houseplant, the best known is gorgeous purple shamrock *(Oxalis triangularis)* though there are many other beautiful varieties of varying size and colour. Did you know that the leaves of this plant are not only edible, they are delicious. Tasting similar to sour cherry or apple, they add beauty and a surprising punch to salads.

Aloe vera – The classic aloe is easy to find, and its jelly-like interior is a fantastic topical treatment for dry, irritated skin – the ultimate after-sun! However, did you know it can do just as much good to the inside of you? Add a chunk of the leaf interior to your fruit smoothie to enjoy a soothing effect all the way through your digestive tract.

Canna lilies – Cannas are stunning foliage plants in many eye-popping colours from deepest purple to rainbow pin-stripes. Also known as Indian shot or arrow root, it has been cultivated for thousands of years for its edible rhizomes and was a staple food for the Incas. Tubers can be eaten raw and cooked, and food can be wrapped and baked in the leaves to add extra flavour to dishes. This is a fast-growing plant that would usually need regular potting on, instead take the opportunity to harvest the outer rhizomes for food and replace the remaining ones to continue to grow and divide.

Agave – This stunning architectural plant is a staple for succulent fans and some species can grow huge! It also offers a sweet nectar used as an alternative to sugar and is the base ingredient in tequila. The flowers are edible if boiled or roasted and a sweet syrup can be made from boiling the leaves and the flower stalks. Some agaves offer other edible parts so research online what specific specimens have to offer.

Banana – The commonest banana houseplant is the 'Dwarf Cavendish' and despite claims it is inedible, a happy plant will produce a crop of small and perfectly edible little bananas. They are not like the big sweet fruit you buy from the market but if used in cooking more like plantain, they can offer an exciting new ingredient to play with. For a more classic banana experience make the effort to find a 'Dwarf Orinoco' or 'Dwarf Brazilian' plant.

Prickly pear – This easy-to-grow cactus provides the classic pad silhouette and offers large crops of sweet fruit. Don't forget to wear gloves while picking the fruit and carefully brush off any spines, though spineless varieties are also available.

Orchids – Vanilla is the best-known edible orchid with its super-flavoured flowers and pods (beans), and it's an exciting addition to any edible jungle, but unless you are an orchid enthusiast you may not know that all orchid flowers are edible. They have a flavour similar to leafy green vegetables while offering a lot more visual excitement than your average spinach.

Go full jungle!

The warmth of your home offers the perfect opportunity to experiment with exciting plants. In my indoor jungle I have mango, lychee, tropical passion fruit, pineapple, vanilla and coffee and I have plans to add pandan and cocoa – yes chocolate! All of these are stunning plants and with the right care can offer exciting crops. One of the easiest to grow, most visually striking and instantly rewarding tropical plants is the dragon fruit cactus. A parasitic plant from tropical Central America, dragon fruit is an amazing climber so be sure to give it something to scale. Once established, it will produce arguably one of the largest (25 cm/10 in), most stunning and most scented flowers of the natural world. It blooms at night, each flower lasting until morning, as in the wild it is pollinated by bats and moths. If pollination is successful, and this may require a bit of help (without a bat), blooms are followed by one of the brightest colored fruits. Usually sold as cuttings, it's a fast grower and can flower and set fruit in its first year, providing instant gratification, a lot of fun with training and quickly creating a real statement. There are two things to bear in mind with dragon fruit. Firstly, this cactus climbs using adventitious roots that will try to penetrate any surface it is trying to scale, so I gave my dragon fruit cuttings a tall, twisted mangrove root to climb. Secondly, cross-pollination provides the best chance of getting fruit, so two plants are better than one. Certain pairings of cultivars pollinate each other well and a good supplier will recommend these: I chose to pair *Hylocereus undatus* (the most commonly cultivated, red-skinned, white-fleshed fruit) with *Hylocereus purpusii* (a bright, purple-fleshed fruit) which is an excellent pollinator. This plant can be pruned hard to create any growth habit you want from bushy to linear to multi-stemmed, and each cutting, when placed in potting compost, will quickly root forming a new plant. Prepare to be popular with friends and family!

Climbers

Who needs art when you have plants? Climbing plants offer the opportunity to fill a wall with gorgeous green. Create your own bespoke trellis using the Indoor Trellis project (see page 123) or take advantage of natural structures in your home. How about sending a vine to wind its way up your stair banisters? Passionflowers are a fabulous choice for the houseplant grower and yes, the orange fruit of the ornamental passionflower is edible and will thrive in the warmth of your home. Kiwi fruit and kiwi berries are another fabulous and fast-growing climber offering lush, fuzzy foliage and heavy crops of fruit once established. For the ultimate in edible and beautiful why not try mashua, a perennial climbing nasturtium originating from South America? Every part of this incredible plant is edible, from the techno-colored tubers, to the peppery leaves, to the abundant trumpet-like flowers.

MAINTENANCE

Don't adopt a one-size-fits-all watering schedule as each houseplant is different! Feed and water as required when the soil has dried out and stick a finger into the compost to see what's really going on! If light levels are not strong enough fix up some grow lights, even adding a daylight bulb into existing lamps to give your plant babies a boost.

Fresh take: edible alternatives to your favourite houseplants

If you love these...	Why not try these...
Devil's ivy *(Pothos)*	Vanilla orchid
Swiss cheese plant *(Monstera)*	Taro *(Colocasia)*
Spider plant *(Chlorophytum)*	Pandan
Bromeliads	Pineapple
Cactus	Prickly pear
Ficus	Citrus
Cast iron plant *(Aspidistra elatior)*	Canna lily
Fiddle leaf fig *(Ficus lyrata)*	Loquat
Snake plant *(Dracaena trifasciata)*	Aloe vera
Succulents	Agave
Euphorbia candelabrum	Dragon fruit *(Hylocereus undatus)*

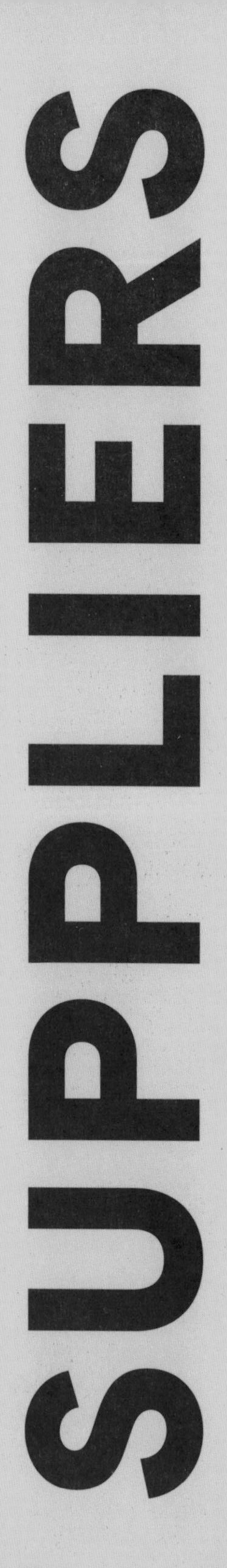

Page 27
FEEDING

Natural Grower Natural Liquid Fertiliser
naturalgrower.co.uk
Contact directly for shipping outside UK.

—

Page 28
WATERING

Warley Fall Green watering can
haws.co.uk
UK and European shipping. Available through third party suppliers in US and Australia.

—

Page 31
TROUBLE SHOOTING

BASF Nemasys Natural fruit and veg protection
Available globally from third party suppliers.

Predatory Mites
Available to purchase online through specialist organic pest control suppliers.

—

Page 47
CLONING SUPERMARKET HERBS

Propagation Vases
amazon.co.uk/shop/shegrowsveg

—

Page 51
SELF-WATERING POTS

Begärlig Vase
ikea.com
Available globally.

—

Page 55
SALAD TROLLEY

Nissafors trolley
ikea.com
Available globally.

Boysenbär plant pot
ikea.com
Available globally.

Gold scissors
amazon.co.uk/shop/shegrowsveg

—

Page 61
MICROGREENS

Window Garden microgreen bundle
windowgarden.us
Shipping to USA and UK.

Gricol shower shelf
amazon.co.uk/shop/shegrowsveg

Ostbit bamboo tray
ikea.com
Available globally.

—

Page 69
INDOOR SPICE GARDEN

Sichuan Pepper
pennardplants.com
UK shipping only. Plants available to source for other suppliers globally.

—

Page 75
HANGING HERB GARDEN

Over door kitchen storage rack
amazon.co.uk/shop/shegrowsveg

Daidai brass colour plant pot
ikea.com
Available globally.

Socker galvanised plant pot
ikea.com
Available globally.

Sirius greenhouse
vitavia.co.uk
UK only.

—

Page 79
ROLLING ROOM DIVIDER

Songmics Clothes rail
amazon.co.uk/shop/shegrowsveg

Vindfläkt planter
ikea.com
Available globally.

Page 93
KOKEDAMA

Citrus trees
suttons.co.uk
UK shipping only. Plants available to source for other suppliers globally.

—

Page 99
EDIBLE LIVING WALL

Wonderwall Wall Planters
wonderwall-promo.co.uk

—

Page 103
HYDROPONICS

IGWorks iHarvest indoor garden
igworks.com
Shipping to USA and Europe. Not currently available in Australia.

Click and Grow smart indoor garden
clickandgrow.com
Shipping to Europe and North USA. Available in Australia through third party suppliers.

Botanium smart pot
botanium.se
Global shipping.

Acqua garden
acquatower.com
Global shipping.

—

Page 109
FARM DESK

Omar shelving system
ikea.com
Available globally.

Kungfors hanging cutlery pot
ikea.com
Available globally.

LED Grow lights
amazon.co.uk/shop/shegrowsveg

Boysenbär plant pot
ikea.com
Available globally.

Ostbit bamboo tray
ikea.com
Available globally.

—

Page 115
EDIBLE TERRARIUM

Fabrikör display cabinet
ikea.com
Available globally.

Vanilla orchid
canerius.com
Global shipping.

—

Page 123
INDOOR TRELLIS

Wall planter
gisellagraham.co.uk
Available globally through third party suppliers.

—

Page 127
CUTLERY POT VERTICAL VEG GARDEN

Fintorp range
ikea.com
Available globally.

Page 133
INDOOR GREENHOUSE

Milbo display cabinet
ikea.com
Available globally.

Fabrikör display cabinet
ikea.com
Available globally.

—

Page 137
EDIBLE ART

Bespoke box frames
pictureframesexpress.co.uk
Shipping to UK and Europe. Other online made to order frames services are available globally.

Creeping herb plants
urbanherbs.co.uk
UK shipping only.

—

Page 143
EAT YOUR HOUSEPLANTS

Artisan wall planter
horticusliving.com
Global shipping available on request.

Speciality tropical plants including dragon fruit cuttings
canerius.com
Global shipping.

Acknowledgements

This book is dedicated to my incredible girls Skyla and Elodie who have demonstrated miraculous patience and understanding for two such small people, while their Mummy was so busy. Girls, you are my inspiration and my motivation, it is all for you! Never stop being anything but the amazing little people you are and let this book show you that with hard work and determination, you can be and do anything in the world that you want!

To Mike, my rock, I am beyond grateful for your total unwavering faith and support and your regular pep talks, even while you were in the middle of nowhere in the artic, I couldn't have done any of this without you and there certainly would have been a lot less laughter along the way.

To my superhuman Mum Karen and stepfather Robin, who fully isolated from the world so they could provide a comprehensive lockdown childcare service, this book would never have happened without your help, love and support, thank you from the bottom of my heart.

Thank you to my amazing Dad David and beloved little sis Amy for always believing in me and encouraging me in all the various phases of my life and career iterations. You are both always there for me no matter what, ready at a moment's notice with love, support and sound advice.

Thank you Tessa Cobley from Ladybird Plantcare for sharing your boundless organic pest control knowledge. Thanks to Matthew Soper from Hampshire Carnivorous Plants for taking the time to discuss the pros and cons of plants for pest control. Andy Perry from Urban Herbs, it has been a pleasure to be part of your journey and I am so glad and grateful that you have now also been a part of mine, thank you for your boundless enthusiasm and exceptional plants. Thank you to Rhino Greenhouses for your support. A huge thanks to Suttons for the array of exceptional plants that have helped bring this book to life. Thanks to Ben Poole for coming to my aid, the wood was beautiful. Thanks also Ray McCune from Scaped Nature for a crash course in the finer points of terrarium making, I think you may have started a bit of an addiction.

A huge thank you to Philippa Langley for the incredible photos and Rachel Vere for making sure each and every shot was styled to utter perfection. Thanks to designer Claire Warner for making this book more beautiful than I could possibly have imagined and to Amber Day for the stunning illustrations that have helped bring the projects to life.

Finally, a huge thank you to the team at Hardie Grant Publishing for seeing the potential in me and giving me this opportunity, and especially to my fabulous editor Eve Marleau for helping me make this book a reality in the midst of a global pandemic that definitely was not part of the plan! I'm so proud of what we have created here.

About the author

Lucy Hutchings is an ex Couture jewellery designer turned grow your own Instagramer and writer. She lives in rural Suffolk, UK, with her partner, explorer, chef and fellow writer Mike Keen, and her two daughters Skyla and Elodie. Lucy's uses her background as a designer to show how inspiring growing your own food can be, and a myriad diverse and beautiful ways in which it can be grown, even in small spaces, and shares this with her loyal Instagram followers (@shegrowsveg).

As a passionate advocate for food sustainability, Lucy started her blog www.shegrowsveg.com to catalogue and share her journey of discovery through the subject with weekly posts on everything from futureproof farming, seasonal and local eating, wild and foraged food and of course inspiration for growing your own.

When she isn't writing she can usually be found outside, embracing her love of nature, usually with camera in hand.

INDEX

Published in 2021 by Hardie Grant Books, an imprint of Hardie Grant Publishing

Hardie Grant Books (London)
5th & 6th Floors
52–54 Southwark Street
London SE1 1UN

Hardie Grant Books (Melbourne)
Building 1, 658 Church Street
Richmond, Victoria 3121

hardiegrantbooks.com

British Library Cataloguing-in-Publication Data. A catalogue record for this book is available from the British Library.

Get Up and Grow by Lucy Hutchings

ISBN: 978-1-78488-392-8

Publisher: Kajal Mistry
Commissioning Editor: Eve Marleau
Design: Claire Warner Studio
Photography: Philippa Langley
Illustrations: Amber Day
Copy Editor: Helen Griffin
Proofreader: Gilliam Haslam
Indexer: Vanessa Bird
Production Controller: Sinead Hering

Colour Reproduction by p2d
Printed and bound in China
by Leo Paper Products Ltd.